Principles of Waste Heat Recovery

Principles of

Waste Heat Recovery

Robert Goldstick
Albert Thumann, CEM, PE

THE FAIRMONT PRESS, INC.
P.O. Box 14227, Atlanta, Georgia 30324

Library of Congress Cataloging-in-Publication Data

Goldstick, Robert, 1943-
    Principles of waste heat recovery.

    Based on: The waste heat recovery handbook/Robert Goldstick.
    Includes index.
    1. Heat recovery. I. Thuman, Albert. II. Goldstick,
Robert, 1943-   . Waste heat recovery handbook.
III. Title.
TJ260.G57   1986   621.402   85-45876
ISBN 0-88173-015-7

PRINCIPLES OF WASTE HEAT RECOVERY
©1986 by The Fairmont Press, Inc.

While every effort is made to provide dependable information, the publisher and author cannot be held responsible for any errors or omissions.

ISBN 0-88173-015-7

Printed in the United States of America

Preface

Much of this work is based upon our previous authored reference *The Waste Heat Recovery Handbook*. An extensive new chapter, "Recovering Process Waste Heat" has been added. Hopefully this addition with its simplified heat recovery diagrams will give you additional insight on how to economically recapture waste heat.

Contents

Acknowledgments

A wide variety of authorities have contributed to this work. Special appreciation is given to Mr. William F. Kenney who authored Manual 8 of the Industrial Energy Conservation Manual which is included in Chapter 8. Several "classic" government publications and papers presented at the World Energy Engineering Congress sponsored by the Association of Energy Engineers have been incorporated into appropriate chapters. Appreciation is expressed to all those who have contributed to this reference.

1

Waste Heat Recovery Basics

INTRODUCTION

Waste heat is heat which is generated in a process but then "dumped" to the environment even though it could still be reused for some useful and economic purpose.

The essential quality of heat is not the amount but rather its "value."

The strategy of how to recover this heat depends in part on the temperature of the waste heat gases and the economics involved.

This text will present the various methods involved in traditionally recovering waste heat. In addition new technologies such as condensation heat recovery will be discussed. Even though these technologies are relatively new in the United States market, they have been applied in Europe for the last ten years.

Portions of material used in Chapters 1 and 2 are based upon the *Waste Heat Management Guidebooks* published by the U.S. Department of Commerce/National Bureau of Standards. The authors express appreciation to Kenneth G. Kreider; Michael B. McNeil; W. M. Rohrer, Jr.; R. Ruegg; B. Leidy and W. Owens who have contributed extensively to this publication.

SOURCES OF WASTE HEAT

Sources of waste energy can be divided according to temperature into three temperature ranges. The high temperature range refers to temperatures above 1200F. The medium temperature range is between 450F and 1200F, and the low temperature range is below 450F.

High and medium temperature waste heat can be used to produce process steam. If one has high temperature waste heat, instead of producing steam directly, one should consider the possibility of using the high temperature energy to do useful work before the waste heat is extracted. Both gas and steam turbines are useful and fully developed heat engines.

In the low temperature range, waste energy which would be otherwise useless can sometimes be made useful by application of mechanical work through a device called the heat pump.

HIGH TEMPERATURE HEAT RECOVERY

The combustion of hydrocarbon fuels produces product gases in the high temperature range. The maximum theoretical temperature possible in atmospheric combustors is somewhat under 3500F, while measured flame temperatures in practical combustors are just under 3000F. Secondary air or some other dilutant is often admitted to the combustor to lower the temperature of the products to the required process temperature, for example to protect equipment, thus lowering the practical waste heat temperature.

Table 1-1 gives temperatures of waste gases from industrial process equipment in the high temperature range. All of these result from direct fuel fired processes.

Table 1-1

Type of Device	Temperature F
Nickel refining furnace	2500–3000
Aluminum refining furnace	1200–1400
Zinc refining furnace	1400–2000
Copper refining furnace	1400–1500
Steel heating furnaces	1700–1900
Copper reverberatory furnace	1650–2000
Open hearth furnace	1200–1300
Cement kiln (Dry process)	1150–1350
Glass melting furnace	1800–2800
Hydrogen plants	1200–1800
Solid waste incinerators	1200–1800
Fume incinerators	1200–2600

MEDIUM TEMPERATURE HEAT RECOVERY

Table 1-2 gives the temperatures of waste gases from process equipment in the medium temperature range. Most of the waste heat in this temperature range comes from the exhausts of directly fired process units. Medium temperature waste heat is still hot enough to allow consideration of the extraction of mechanical work from the waste heat, by a steam or gas turbine. Gas turbines can be economically utilized in some cases at inlet pressures in the range of 15 to 30 lb/in^{2}g. Steam can be generated at almost any desired pressure and steam turbines used when economical.

Table 1-2

Type of Device	Temperature F
Steam boiler exhausts	450–900
Gas turbine exhausts	700–1000
Reciprocating engine exhausts	600–1100
Reciprocating engine exhausts (turbocharged)	450–700
Heat treating furnaces	800–1200
Drying and baking ovens	450–1100
Catalytic crackers	800–1200
Annealing furnace cooling systems	800–1200

LOW TEMPERATURE HEAT RECOVERY

Table 1-3 lists some heat sources in the low temperature range. In this range it is usually not practicable to extract work from the source, though steam production may not be completely excluded if there is a need for low pressure steam. Low temperature waste heat may be useful in a supplementary way for preheating purposes. Taking a common example, it is possible to use economically the energy from an air conditioning condenser operating at around 90F to heat the domestic water supply. Since the hot water must be heated to about 160F, obviously the air conditioner waste heat is not hot enough. However, since the cold water enters the domestic water system at about 50F, energy interchange can take place raising the water to something less than 90F. Depending upon the relative air conditioning lead and hot water requirements, any excess condenser heat

Table 1-3

Source	Temperature F
Process steam condensate	130–190
Cooling water from:	
Furnace doors	90–130
Bearings	90–190
Welding machines	90–190
Injection molding machines	90–190
Annealing furnaces	150–450
Forming dies	80–190
Air compressors	80–120
Pumps	80–190
Internal combustion engines	150–250
Air conditioning and	
refrigeration condensers	90–110
Liquid still condensers	90–190
Drying, baking and curing ovens	200–450
Hot processed liquids	90–450
Hot processed solids	200–450

can be rejected and the additional energy required by the hot water provided by the usual electrical or fired heater.

WASTE HEAT RECOVERY APPLICATIONS

To use waste heat from sources such as those above, one often wishes to transfer the heat in one fluid stream to another (e.g., from flue gas to feedwater or combustion air). The device which accomplishes the transfer is called a heat exchanger. In the discussion immediately below is a listing of common uses for waste heat energy and in some cases, the name of the heat exchanger that would normally be applied in each particular case.

The equipment that is used to recover waste heat can range from something as simple as a pipe or duct to something as complex as a waste heat boiler.

Some applications of waste heat are as follows:

• Medium to high temperature exhaust gases can be used to preheat the combustion air for:
Boilers using air-preheaters
Furnaces using recuperators

Ovens using recuperators

Gas turbines using regenerators

• Low to medium temperature exhaust gases can be used to preheat boiler feedwater or boiler makeup water using *economizers,* which are simply gas-to-liquid water heating devices.

• Exhaust gases and cooling water from condensers can be used to preheat liquid and/or solid feedstocks in industrial processes. Finned tubes and tube-in-shell *heat exchangers* are used.

• Exhaust gases can be used to generate steam in *waste heat boilers* to produce electrical power, mechanical power, process steam, and any combination of above.

• Waste heat may be transferred to liquid or gaseous process units directly through pipes and ducts or indirectly through a secondary fluid such as steam or oil.

• Waste heat may be transferred to an intermediate fluid by heat exchangers or waste heat boilers, or it may be used by circulating the hot exit gas through pipes or ducts. Waste heat can be used to operate an absorption cooling unit for air conditioning or refrigeration.

THE WASTE HEAT RECOVERY SURVEY

In order to identify source of waste heat a survey is usually made. Figure 1-1 illustrates a survey form which can be used for the Waste Heat Audit. It is important to record flow and temperature of waste gases.

Composition data is required for heat recovery and system design calculations. Be sure to note contaminants since this factor could limit the type of heat recovery equipment to apply. Contaminants can foul or plug heat exchangers.

Operation schedule affects the economics and type of equipment to be specified. For example, an incinerator that is only used one shift per day may require a different method of recovering discharges than if it were used three shifts a day. A heat exchanger used for waste heat recovery in this service would soon deteriorate due to metal fatigue. A different type of heat recovery incinerator utilizing heat storage materials such as rock or ceramic would be more suitable.

SURVEY FORM FOR INDUSTRIAL PROCESS UNITS

NAME OF PROCESS UNIT ______________ INVENTORY NUMBER ______________

LOCATION OF PROCESS UNIT, PLANT NAME ______________ BUILDING ______________

MANUFACTURER ______________ MODEL ______ SERIAL NUMBER ______________

	NAME	FIRING RATE	HHV	TEMPERATURE OF			FLUE GAS COMPOSITION % VOLUME				
				COMB. AIR	FUEL	STACK	CO_2	O_2	CO	CH	N_2
PRIMARY FUEL											
FIRST ALTERNAT.											
SECOND ALTERNAT.											

	FLOW PATH 1	FLOW PATH 2	FLOW PATH 3	FLOW PATH 4
FLUID COMPOSITION				
FLOW RATE				
INLET TEMPERATURE				
OUTLET TEMPERATURE				
DESCRIPTION				

ANNUAL HOURS OPERATION ______________ ANNUAL CAPACITY FACTOR, % ______________

ANNUAL FUEL CONSUMPTION: PRIMARY FUEL ______; FIRST ALTERN. ______ SEC. ALTERN. ______

PRESENT FUEL COST: PRIMARY FUEL ______; FIRST ALTERN. ______ SEC. ALTERN. ______

ANNUAL ELECTRICAL ENERGY CONSUMPTION, KWHR. ______________

PRESENT ELECTRICAL ENERGY RATE ______________

CONTAMINANTS ______________

Figure 1-1. Waste Heat Survey

WASTE HEAT RECOVERY CALCULATIONS

Heat Balance

Waste heat recovery calculations usually start with the heat balance. A heat balance is an analysis of a process which shows where all the heat comes from and where it goes. This is a vital tool in assessing the profit implications of heat losses and proposed waste heat utilization projects. The heat balance for a steam boiler, process furnace, air conditioner, etc., must be derived from measurements made during actual operating periods. The measurements that are needed to get a complete heat balance involve: energy inputs, energy losses to the environment, and energy discharges.

Energy Input

Energy enters most process equipment either as chemical energy in the form of fossil fuels, of sensible enthalpy of fluid streams, of latent heat in vapor streams, or as electrical energy.

For each input it is necessary to meter the quantity of fluid flowing or the electrical current. This means that if accurate results are to be obtained, submetering for each flow is required (unless all other equipment served by a main meter can be shut down so that the main meter can be used to measure the inlet flow to the unit). It is not necessary to continuously submeter every flow since temporary installations can provide sufficient information. In the case of furnaces and boilers that use pressure ratio combustion controls, the control flow meters can be utilized to yield the correct information. It should also be pointed out that for furnaces and boilers only the fuel need be metered. Tests of the exhaust products provide sufficient information to derive the oxidant (usually air) flow if accurate fuel flow data are available.

For electrical energy inflows, the current is measured with an ammeter, or a kilowatt-hour meter may be installed as a submeter. Ammeters using split core transformers are available for measuring alternating current flow without opening the line. These are particularly convenient for temporary installations.

In addition to measuring the flow for each inlet stream, it is necessary to know the chemical composition of the stream. For air, water, and other pure substances no tests for composition are re-quited, but for fossil fuels the composition must be determined by chemical analysis or secured from the fuel supplier. For vapors one should know the quality—this is the mass fraction of vapor present in the mixture of vapor and droplets. Measurement of quality is made with a vapor calorimeter which requires only a small sample of the vapor stream.

Other measurements that are required are the entering temper-atures of the inlet stream of fluid and the voltages of the electrical energy entering (unless kilowatt-hour meters are used).

The testing routines discussed above involve a good deal of time, trouble, and expense. However, they are necessary for accurate analyses and may constitute the critical element in the engineering and economic analyses required to support decisions to expend cap-ital on waste heat recovery equipment.

From the heat balance, the heat recovered from the source is determined by Formula 1-1.

$$q = m \, c_p \, \Delta T \qquad \qquad 1\text{-}1$$

where: q = heat recovered, Btu/hr
m = mass flow rate lbs/hr
c_p = specific heat of fluid, Btu/lb°F
ΔT = temperature change of gas or liquid during heat recov-ery °F

If the flow is air, then Formula 1-1 can be expressed as

$$q = 1.08 \text{ CFM } \Delta T \qquad \qquad 1\text{-}2$$

where: CFM = volume flow rate in standard CFM

If the flow is water, then Formula 1-1 can be expressed as

$$q = 500 \text{ GPM } \Delta T \qquad \qquad 1\text{-}3$$

where: GPM = volume flow rate in gallons per minute

SIM 1-1

A waste heat audit survey indicates 10,000 lb/hr of water at 190° is discharged to the sewer. How much heat can be saved by utilizing this fluid as makeup to the boiler instead of the 70°F feed-water supply? Fuel cost is $6 per million Btu, boiler efficiency .8, and hours of operation 4000.

Analysis

$$q = mc_p \quad \Delta T = 10,000 \times (190-70) = 1.2 \times 10^6 \text{ Btu/hr}$$
$$\text{Savings} = 1.2 \times 10^6 \times 4000 \times \$6/10^6/.8 = \$36,000$$

Heat Transfer by Convection

Convection is the transfer of heat between a fluid, gas, or liquid. Formula 1-4 is indicative of the basic form of convective heat transfer. U_0, in this case, represents the convection film conductance, Btu/ft^2 · hr · °F.

Heat transferred for heat exchanger applications is predominantly a combination of conduction and convection expressed as:

$$q = U_0 A \Delta T_m \qquad\qquad 1\text{-}4$$

where: q = rate of heat flow by convection, Btu/hr
U_0 = is the overall heat transfer coefficient Btu/ft^2 · hr · °F
A = is the area of the tubes in square feet
ΔT_m = is the logarithmic mean temperature difference and represents the situation where the temperature of two fluids change as they transverse the surface.

$$\Delta T_m = \frac{\Delta T_1 - \Delta T_2}{\text{Log}_e [\Delta T_1/\Delta T_2]} \qquad\qquad 1\text{-}5$$

To understand the different logarithmic mean temperature relationships, Figure 1-2 should be used. Referring to Figure 1-2, the ΔT_m for the counterflow heat exchanger is:

$$\Delta T_m = \frac{(t_1 - t_2') - (t_2 - t_1')}{\text{Log}_e [t_1 - t_2'/t_2 - t_1']} \qquad\qquad 1\text{-}6$$

The ΔT_m for the parallel flow heat exchanger is:

$$\Delta T_m = \frac{(t_1 - t_1') - (t_2 - t_2')}{\text{Log}_e [t_1 - t_1'/t_2 - t_2']} \qquad\qquad 1\text{-}7$$

A. COUNTERFLOW

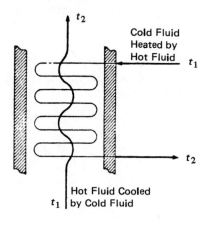

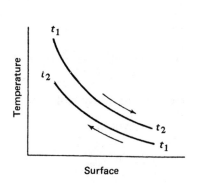

B. PARALLEL FLOW

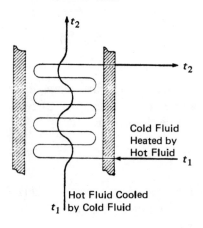

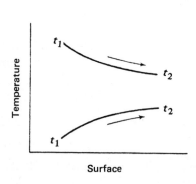

Figure 1-2. Temperature Relationships for Heat Exchangers

WASTE HEAT RECOVERY EQUIPMENT

Industrial heat exchangers have many pseudonyms. They are sometimes called recuperators, regenerators, waste heat steam generators, condensers, heat wheels, temperature and moisture exchangers, etc. Whatever name they may have, they all perform one basic function: the transfer of heat.

Heat exchangers are characterized as single or multipass gas to gas, liquid to gas, liquid to liquid, evaporator, condenser, parallel flow, counterflow, or crossflow. The terms single or multipass refer to the heating or cooling media passing over the heat transfer surface once or a number of times. Multipass flow involves the use of internal baffles. The next three terms refer to the two fluids between which heat is transferred in the heat exchanger, and imply that no phase changes occur in those fluids. Here the term "fluid" is used in the most general sense. Thus, we can say that these terms apply to non-evaporator and noncondensing heat exchangers. The term evaporator applies to a heat exchanger in which heat is transferred to an evaporating (boiling) liquid, while a condenser is a heat exchanger in which heat is removed from a condensing vapor. A parallel flow heat exchanger is one in which both fluids flow in approximately the same direction whereas in counterflow the two fluids move in opposite directions. When the two fluids move at right angles to each other, the heat exchanger is considered to be of the crossflow type.

The principal methods of reclaiming waste heat in industrial plants make use of heat exchangers. The heat exchanger is a system which separates the stream containing waste heat and the medium which is to absorb it, but allows the flow of heat across the separation boundaries. The reasons for separating the two streams may be any of the following:

(1) A pressure difference may exist between the two streams of fluid. The rigid boundaries of the heat exchanger can be designed to withstand the pressure difference.

(2) In many, if not most, cases the one stream would contaminate the other, if they were permitted to mix. The heat exchanger prevents mixing.

(3) Heat exchangers permit the use of an intermediate fluid

better suited than either of the principal exchange media for transporting waste heat through long distances. The secondary fluid is often steam, but another substance may be selected for special properties.

(4) Certain types of heat exchangers, specifically the heat wheel, are capable of transferring liquids as well as heat. Vapors being cooled in the gases are condensed in the wheel and later re-evaporated into the gas being heated. This can result in improved humidity and/or process control, abatement of atmospheric air pollution, and conservation of valuable resources.

The various names or designations applied to heat exchangers are partly an attempt to describe their function and partly the result of tradition within certain industries. For example, a recuperator is a heat exchanger which recovers waste heat from the exhaust gases of a furnace to heat the incoming air for combustion. This is the name used in both the steel and the glass making industries. The heat exchanger performing the same function in the steam generator of an electric power plant is termed an air preheater, and in the case of a gas turbine plant, a regenerator.

However, in the glass and steel industries the word regenerator refers to two chambers of brick checkerwork which alternately absorb heat from the exhaust gases and then give up part of that heat to the incoming air. The flows of flue gas and of air are periodically reversed by valves so that one chamber of the regenerator is being heated by the products of combustion while the other is being cooled by the incoming air. Regenerators are often more expensive to buy and more expensive to maintain than are recuperators, and their application is primarily in glass melt tanks and in open hearth steel furnaces.

It must be pointed out, however, that although their functions are similar, the three heat exchangers mentioned above may be structurally quite different as well as different in their principal modes of heat transfer. A more complete description of the various industrial heat exchangers follows later in this chapter and details of their differences will be clarified.

The specification of an industrial heat exchanger must include the heat exchange capacity, the temperatures of the fluids, the allowable pressure drop in each fluid path, and the properties and volumetric flow of the fluids entering the exchanger. These specifications will determine construction parameters and thus the cost of the heat exchanger. The final design will be a compromise between pressure drop, heat exchanger effectiveness, and cost. Decisions leading to that final design will balance out the cost of maintenance and operation of the overall system against the fixed costs in such a way as to minimize the total. Advice on selection and design of heat exchangers is available from vendors.

The essential parameters which should be known in order to make an optimum choice of waste heat recovery devices are:

- Temperature of waste heat fluid
- Flow rate of waste heat fluid
- Chemical composition of waste heat fluid
- Minimum allowable temperature of waste heat fluid
- Temperature of heated fluid
- Chemical composition of heated fluid
- Maximum allowable temperature of heated fluid
- Control temperature, if control required

In the rest of this chapter, some common types of waste heat recovery devices are discussed in some detail.

GAS TO GAS HEAT EXCHANGERS

Recuperators

The simplest configuration for a heat exchanger is the metallic radiation recuperator which consists of two concentric lengths of metal tubing as shown in Figure 1-3.

The inner tube carries the hot exhaust gases while the external annulus carries the combustion air from the atmosphere to the air inlets of the furnace burners. The hot gases are cooled by the incoming combustion air which now carries additional energy into the combustion chamber. This is energy which does not have to be supplied by the fuel; consequently, less fuel is burned for a given furnace loading. The saving in fuel also means a decrease in combustion air and therefore stack losses are decreased not only by lowering the

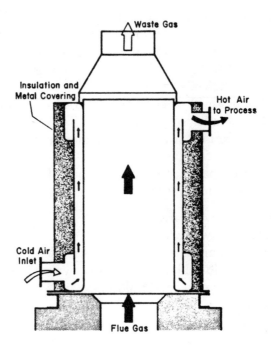

Figure 1-3. Diagram of Metallic Radiation Recuperator

stack gas temperatures, but also by discharging smaller quantities of exhaust gas. This particular recuperator gets its name from the fact that a substantial portion of the heat transfer from the hot gases to the surface of the inner tube take place by radiative heat transfer. The cold air in the annulus, however, is almost transparent to infrared radiation so that only convection heat transfer takes place to the incoming air. As shown in the diagram, the two gas flows are usually parallel, although the configuration would be simpler and the heat transfer more efficient if the flows were opposed in direction (or counterflow). The reason for the use of parallel flow is that recuperators frequently serve the additional function of cooling the duct carrying away the exhaust gases, and consequently extending its service life.

The inner tube is often fabricated from high temperature materials such as stainless steels of high nickel content. The large temperature differential at the inlet causes differential expansion, since the outer shell is usually of a different and less expensive material. The mechanical design must take this effect into account. More elaborate designs of radiation recuperators incorporate two sections; the bottom operating in parallel flow and the upper section using the more efficient counterflow arrangement. Because of the large axial expansions experienced and the stress conditions at the bottom of the recuperator, the unit is often supported at the top by a free standing support frame with an expansion joint between the furnace and recuperator.

A second common configuration for recuperators is called the tube type or convective recuperator. As seen in the schematic diagram of Figure 1-4, the hot gases are carried through a number of parallel small diameter tubes, while the incoming air to be heated enters a shell surrounding the tubes and passes over the hot tubes one or more times in a direction normal to their axes.

If the tubes are baffled to allow the gas to pass over them twice, the heat exchanger is termed a two-pass recuperator; if two baffles are used, a three-pass recuperator, etc. Although baffling increases both the cost of the exchanger and the pressure drop in the combustion air path, it increases the effectiveness of heat exchange. Shell-and tube-type recuperators are generally more compact and have a

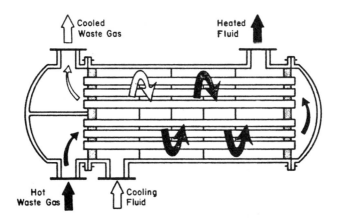

Figure 1-4. Diagram of Convective-Type Recuperator

higher effectiveness than radiation recuperators, because of the larger heat transfer area made possible through the use of multiple tubes and multiple passes of the gases.

The principal limitation on the heat recovery of metal recuperators is the reduced life of the liner at inlet temperatures exceeding 2000F. At this temperature, it is necessary to use the less efficient arrangement of parallel flows of exhaust gas and coolant in order to maintain sufficient cooling of the inner shell. In addition, when furnace combustion air flow is dropped back because of reduced load, the heat transfer rate from hot waste gases to preheat combustion air becomes excessive, causing rapid surface deterioration. Then, it is usually necessary to provide an ambient air by-pass to cool the exhaust gases.

In order to overcome the temperature limitations of metal recuperators, ceramic tube recuperators have been developed, whose materials allow operation on the gas side to 2800F and on the preheated air side to 2200F on an experimental basis, and to 1500F on a more or less practical basis. Early ceramic recuperators were built of tile and joined with furnace cement, and thermal cycling caused cracking of joints and rapid deterioration of the tubes. Later developments introduced various kinds of short silicon carbide tubes which can be joined by flexible seals located in the air headers. This kind of patented design illustrated in Figure 1-5 maintains the seals at comparatively low temperatures and has reduced the seal leakage rates to a few percent.

Earlier designs had experienced leakage rates from 8 to 60 percent. The new designs are reported to last two years with air preheat temperatures as high as 1300F, with much lower leakage rates.

An alternative arrangement for the convective type recuperator, in which the cold combustion air is heated in a bank of parallel vertical tubes which extend into the flue gas stream, is shown schematically in Figure 1-6. The advantage claimed for this arrangement is the ease of replacing individual tubes, which can be done during full capacity furnace operation. This minimizes the cost, the inconvenience, and possible furnace damage due to a shutdown forced by recuperator failure.

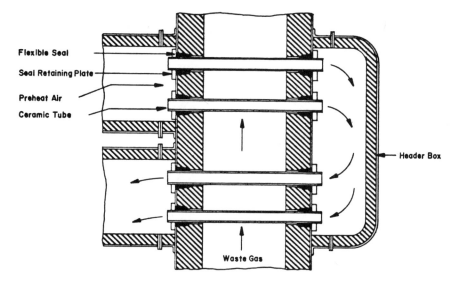

Figure 1-5. Ceramic Recuperator

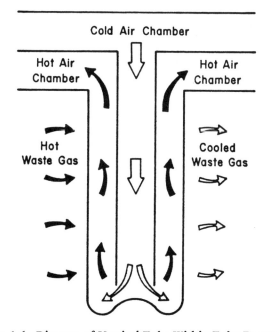

Figure 1-6. Diagram of Vertical Tube-Within-Tube Recuperator

For maximum effectiveness of heat transfer, combinations of radiation type and convective type recuperators are used, with the convective type always following the high temperature radiation recuperator. A schematic diagram of this arrangement is seen in Figure 1-7.

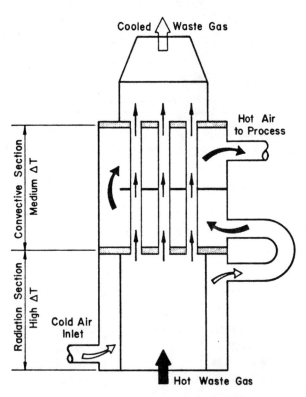

Figure 1-7. Diagram of Combined Radiation and Convective Type Recuperator

Although the use of recuperators conserves fuel in industrial furnaces, and although their original cost is relatively modest, the purchase of the unit is often just the beginning of a somewhat more extensive capital improvement program. The use of a recuperator, which raises the temperature of the incoming combustion air, may require purchase of high temperature burners, larger diameter air lines with flexible fittings to allow for expansion, cold air lines for cooling the burners, modified combustion controls to maintain the required air/fuel ratio despite variable recuperator heating, stack dampers, cold air bleeds, controls to protect the recuperator during blower failure or power failures, and larger fans to overcome the additional pressure drop in the recuperator. It is vitally important to protect the recuperator against damage due to excessive temperatures, since the cost of rebuilding a damaged recuperator may be as high as 90 percent of the initial cost of manufacture and the drop in efficiency of a damaged recuperator may easily increase fuel costs by 10 to 15 percent.

Figure 1-8 shows a schematic diagram of one radiant tube burner fitted with a radiation recuperator. With such a short stack, it is necessary to use two annuli for the incoming air to achieve reasonable heat exchange efficiencies.

Recuperators are used for recovering heat from exhaust gases to heat other gases in the medium to high temperature range. Some typical applications are in soaking ovens, annealing ovens, melting furnaces, afterburners and gas incinerators, radiant-tube burners, reheat furnaces, and other gas to gas waste heat recovery applications in the medium to high temperature range.

Heat Wheels

A rotary regenerator (also called an air preheater or a heat wheel) is finding increasing applications in low to medium temperature waste heat recovery. Figure 1-9 is a sketch illustrating the application of a heat wheel. It is a sizable porous disk, fabricated from some material having a fairly high heat capacity, which rotates between two side-by-side ducts; one a cold gas duct, the other a hot gas duct. The axis of the disk is located parallel to, and on the partition between the two ducts. As the disk slowly rotates, sensible heat (and in some cases, moisture containing latent heat) is transferred to the disk by the hot air and as the disk rotates, from the disk to the cold air. The overall efficiency of sensible heat transfer for this kind of regenerator can be as high as 85 percent. Heat wheels have been built as large as 70 feet in diameter with air capacities up to 40,000 ft^3/min. Multiple units can be used in parallel. This may help to prevent

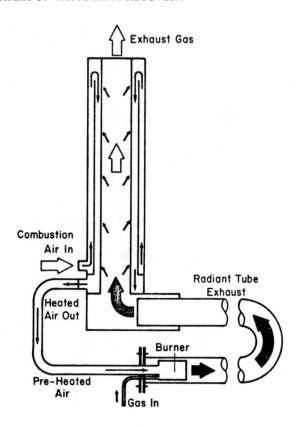

Figure 1-8. Diagram of a Small Radiation-Type Recuperator Fitted to a Radiant Tube Burner

a mismatch between capacity requirements and the limited number of sizes available in packaged units. In very large installations such as those required for preheating combustion air in fixed station electrical generating stations, the units are custom designed.

The limitation on temperature range for the heat wheel is primarily due to mechanical difficulties introduced by uneven expansion of the rotating wheel when the temperature differences mean large differential expansion, causing excessive deformations of the wheel and thus difficulties in maintaining adequate air seals between duct and wheel.

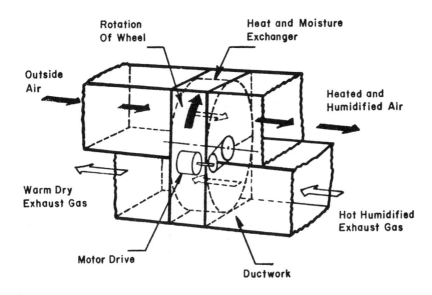

**Figure 1-9. Heat and Moisture Recovery Using a
Heat Wheel Type Regenerator**

Heat wheels are available in four types. The first consists of a metal frame packed with a core of knitted mesh stainless steel or aluminum wire, resembling that found in the common metallic kitchen pot scraper; the second, called a laminar wheel, is fabricated from corrugated metal and is composed of many parallel flow passages; the third variety is also a laminar wheel but is constructed from a ceramic matrix of honeycomb configuration. This type is used for higher temperature applications with a present-day limit of about 1600F. The fourth variety is of laminar construction but the flow passages are coated with a hygroscopic material so that latent heat may be recovered. The packing material of the hygroscopic wheel may be any of a number of materials. The hygroscopic material is often termed a dessicant.

Most industrial stack gases contain water vapor, since water vapor is a product of the combustion of all hydrocarbon fuels and since water is introduced into many industrial processes, and part of the process water evaporates as it is exposed to the hot gas stream. Each pound of water requires approximately 1000 Btu for its evaporation at atmospheric pressure, thus each pound of water vapor leaving in the exit stream will carry 1000 Btu of energy with it. This latent heat may be a substantial fraction of the sensible energy in the exit gas stream. A hygroscopic material is one such as lithium chloride (LiCl) which readily absorbs water vapor. Lithium chloride is a solid which absorbs water to form a hydrate, $LiCl \cdot H_2O$, in which one molecule of lithium chloride combines with one molecule of water. Thus, the ratio of water to lithium chloride in $LiCl \cdot H_2O$ is 3/7 by weight. In a hygroscopic heat wheel, the hot gas stream gives up part of its water vapor to the coating; the cool gases which enter the wheel to be heated are drier than those in the inlet duct and part of the absorbed water is given up to the incoming gas stream. The latent heat of the water adds directly to the total quantity of recovered waste heat. The efficiency of recovery of water vapor can be as high as 50 percent.

Since the pores of heat wheels carry a small amount of gas from the exhaust to the intake duct, cross contamination can result. If this contamination is undesirable, the carryover of exhaust gas can be partially eliminated by the addition of a purge section where a small amount of clean air is blown through the wheel and then exhausted to the atmosphere, thereby clearing the passages of exhaust gas. Figure 1-10 illustrates the features of an installation using a purge section. Note that additional seals are required to separate the purge ducts. Common practice is to use about six air changes of clean air for purging. This limits gas contamination to as little as 0.04 percent and particle contamination to less than 0.2 percent in laminar wheels, and cross contamination to less than 1 percent in packed wheels. If inlet gas temperature is to be held constant, regardless of heating loads and exhaust gas temperatures, then the heat wheel must be driven at variable speed. This requires a variable speed drive and a speed control system using an inlet air temperature sensor as the control element. This feature, however, adds considerably to the cost and complexity of the system. When operating with outside air in periods of high humidity and sub-zero temperatures, heat wheels may require preheat systems to prevent frost formation. When handling gases which contain water-soluble, greasy or adhesive contaminants or large concentrations of process dust, air filters may be required in the exhaust system upstream from the heat wheel.

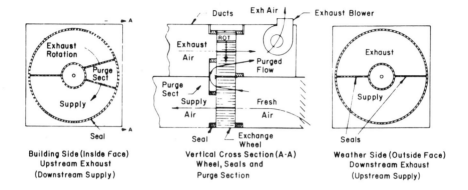

**Figure 1-10. Heat Wheel Equipped with Purge Section to
Clear Contaminants from the Heat Transfer Surface**

One application of heat wheels is in space heating situations where unusually large quantities of ventilation air are required for health or safety reasons. As many as 20 or 30 air changes per hour may be required to remove toxic gases or to prevent the accumulation of explosive mixtures. Comfort heating for that quantity of ventilation air is frequently expensive enough to make the use of heat wheels economical. In the summer season the heat wheel can be used to cool the incoming air from the cold exhaust air, reducing the air conditioning load by as much as 50 percent. It should be pointed out that in many circumstances where large ventilating requirements are mandatory, a better solution than the installation of heat wheels may be the use of local ventilation systems to reduce the hazards and/or the use of infrared comfort heating at principal work areas.

Heat wheels are finding increasing use for process heat recovery in low and moderate temperature environments. Typical applications would be curing or drying ovens and air preheaters in all sizes for industrial and utility boilers.

Air Preheaters

Passive gas to gas regenerators, sometimes called air preheaters, are available for applications which cannot tolerate any cross contamination. They are constructed of alternate channels (see Figure 1-11) which put the flows of the heating and the heated gases in close

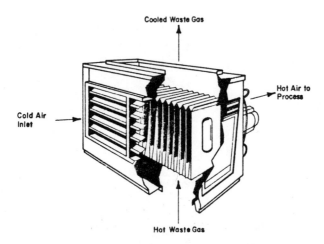

Figure 1-11. A Passive Gas to Gas Regenerator

contact with each other, separated only by a thin wall of conductive metal. They occupy more volume and are more expensive to construct than are heat wheels, since a much greater heat transfer surface area is required for the same efficiency. An advantage, besides the absence of cross-contamination, is the decreased mechanical complexity since no drive mechanism is required. However, it becomes more difficult to achieve temperature control with the passive regeneration and, if this is a requirement, some of the advantages of its basic simplicity are lost.

Gas-to-gas regenerators are used for recovering heat from exhaust gases to heat other gases in the low to medium temperature range. A list of typical applications follows:

- Heat and moisture recovery from building heating and ventilation systems
- Heat and moisture recovery from moist rooms and swimming pools
- Reduction of building air conditioner loads
- Recovery of heat and water from wet industrial processes
- Heat recovery from steam boiler exhaust gases
- Heat recovery from gas and vapor incinerators
- Heat recovery from baking, drying, and curing ovens
- Heat recovery from gas turbine exhausts

- Heat recovery from other gas-to-gas applications in the low through high temperature range.

Heat-Pipe Exchangers

The heat pipe is a heat transfer element that has only recently become commercial, but it shows promise as an industrial waste heat recovery option because of its high efficiency and compact size. In use, it operates as a passive gas-to-gas finned-tube regenerator. As can be seen in Figure 1-12, the elements form a bundle of heat pipes which extend through the exhaust and inlet ducts in a pattern that resembles the structured finned coil heat exchangers. Each pipe, however, is a separate sealed element consisting of an annular wick on the inside of the full length of the tube, in which an appropriate heat transfer fluid is entrained.

Figure 1-13 shows how the heat absorbed from hot exhaust gases evaporates the entrained fluid, causing the vapor to collect in the center core. The latent heat of vaporization is carried in the vapor to the cold end of the heat pipe located in the cold gas duct. Here the vapor condenses giving up its latent heat. The condensed liquid is then carried by capillary (and/or gravity) action back to the hot end where it is recycled. The heat pipe is compact and efficient because: (1) the finned-tube bundle is inherently a good configuration for convective heat transfer in both gas ducts, and (2) the evaporative-condensing cycle within the heat tubes is a highly efficient way of transferring the heat internally. It is also free from cross contamination. Possible applications include:

- Drying, curing and baking ovens
- Waste steam reclamation
- Air preheaters in steam boilers
- Air dryers
- Brick kilns (secondary recovery)
- Reverberatory furnaces (secondary recovery)
- Heating, ventilating and air conditioning systems

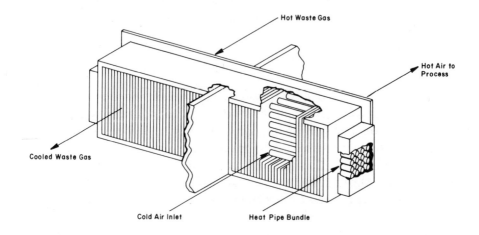

Hot Waste Gas

Hot Air to Process

Cooled Waste Gas

Cold Air Inlet

Heat Pipe Bundle

Figure 1-12. Heat Pipe Bundle Incorporated in Gas to Gas Regenerator

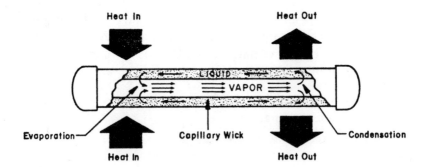

Heat In

Heat Out

LIQUID

VAPOR

Evaporation

Capillary Wick

Condensation

Heat In

Heat Out

Figure 1-13. Heat Pipe Schematic

GAS OR LIQUID TO LIQUID REGENERATORS

Finned-Tube Heat Exchangers

When waste heat in exhaust gases is recovered for heating liquids for purposes such as providing domestic hot water, heating the feedwater for steam boilers, or for hot water space heating, the finned-tube heat exchanger is generally used. Round tubes are connected together in bundles to contain the heated liquid and fins are welded or otherwise attached to the outside of the tubes to provide additional surface area for removing the waste heat in the gases.

Figure 1-14 shows the usual arrangement for the finned-tube exchanger positioned in a duct and details of a typical finned-tube construction. This particular type of application is more commonly known as an economizer. The tubes are often connected all in series but can also be arranged in series-parallel bundles to control the liquid side pressure drop. The air side pressure drop is controlled by the spacing of the tubes and the number of rows of tubes within the duct.

Finned-tube exchangers are available prepackaged in modular sizes or can be made up to custom specifications very rapidly from standard components. Temperature control of the heated liquid is usually provided by a bypass duct arrangement which varies the flow rate of hot gases over the heat exchanger. Materials for the tubes and the fins can be selected to withstand corrosive liquids and/or corrosive exhaust gases.

Finned-tube heat exchangers are used to recover waste heat in the low to medium temperature range from exhaust gases for heating liquids. Typical applications are domestic hot water heating, heating boiler feedwater, hot water space heating, absorption-type refrigeration or air conditioning, and heating process liquids.

Shell and Tube Heat Exchanger

When the medium containing waste heat is a liquid or a vapor which heats another liquid, then the shell and tube heat exchanger must be used since both paths must be sealed to contain the pressures

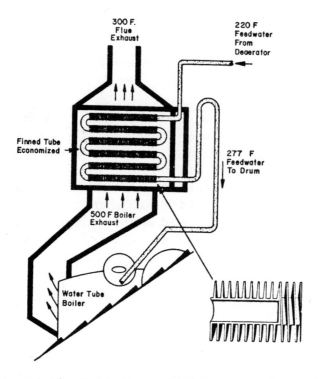

Figure 1-14. Finned-Tube Gas to Liquid Regenerator (Economizer)

of their respective fluids. The shell contains the tube bundle, and usually internal baffles, to direct the fluid in the shell over the tubes in multiple passes. The shell is inherently weaker than the tubes so that the higher pressure fluid is circulated in the tubes while the lower pressure fluid flows through the shell. When a vapor contains the waste heat, it usually condenses, giving up its latent heat to the liquid being heated. In this application, the vapor is almost invariably contained within the shell. If the reverse is attempted, the condensation of vapors within small diameter parallel tubes causes flow instabilities. Tube and shell heat exchangers are available in a wide range of standard sizes with many combinations of materials for the tubes and shells.

Typical applications of shell and tube heat exchangers include heating liquids with the heat contained by condensates from refrigeration and air conditioning systems; condensate from process steam; coolants from furnace doors, grates, and pipe supports; coolants from engines, air compressors, bearings, and lubricants; and the condensates from distillation processes.

Waste Heat Boilers

Waste heat boilers are ordinarily water tube boilers in which the hot exhaust gases from gas turbines, incinerators, etc., pass over a number of parallel tubes containing water. The water is vaporized in the tubes and collected in a steam drum from which it is drawn off for use as heating or processing steam.

Figure 1-15 indicates one arrangement that is used, where the exhaust gases pass over the water tubes twice before they are exhausted to the air. Because the exhaust gases are usually in the medium temperature range and in order to conserve space, a more compact boiler can be produced if the water tubes are finned in order to increase the effective heat transfer area on the gas side. The diagram shows a mud drum, a set of tubes over which the hot gases make a double pass, and a steam drum which collects the steam generated above the water surface. The pressure at which the steam is generated and the rate of steam production depend on the temperature of the hot gases entering the boiler, the flow rate of the hot gases, and the efficiency of the boiler. The pressure of a pure vapor in the presence of its liquid is a function of the temperature of the liquid from which it is evaporated. The steam tables tabulate this relationship between saturation pressure and temperature. Should the waste heat in the exhaust gases be insufficient for generating the required amount of process steam, it is sometimes possible to add auxiliary burners which burn fuel in the waste heat boiler or to add an afterburner to the exhaust gas duct just ahead of the boiler. Waste heat boilers are built in capacities from less than a thousand to almost a million ft^3/min. of exhaust gas.

Typical applications of waste heat boilers are to recover energy from the exhausts of gas turbines, reciprocating engines, incinerators, and furnaces.

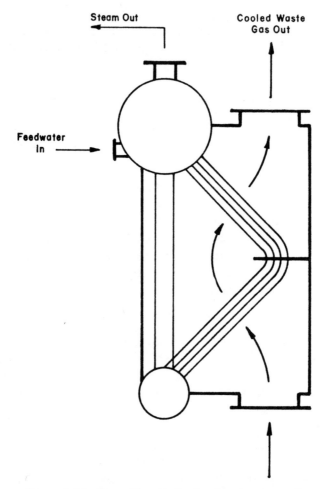

**Figure 1-15. Waste Heat Boiler for Heat Recovery from
Gas Turbines or Incinerators**

Gas and Vapor Expanders

Industrial steam and gas turbines are in an advanced state of development and readily available on a commercial basis. Recently special gas turbine designs for low pressure waste gases have become available; for example, a turbine is available for operation from the top gases of a blast furnace. In this case, as much as 20 MW of power could be generated, representing a recovery of 20 to 30 percent of the available energy of the furnace exhaust gas stream. Maximum top pressures are of the order of 40 lb/in² g.

Perhaps of greater applicability than the last example are steam turbines used for producing mechanical work or for driving electrical generators. After removing the necessary energy for doing work, the steam turbine exhausts partially spent steam at a lower pressure than the inlet pressure. The energy in the turbine exhaust stream can then be used for process heat in the usual ways. Steam turbines are classified as back-pressure turbines, available with allowable exit pressure operation above 400 lb/in² g, or condensing turbines which operate below atmospheric exit pressures. The steam used for driving the turbines can be generated in direct fired or waste heat boilers. A list of typical applications for gas and vapor expanders follows:

- Electrical power generation
- Compressor drives
- Pump drives
- Fan drives

Heat Pumps

In the commercial options previously discussed in this chapter, we find waste heat being transferred from a hot fluid to a fluid at a lower temperature. Heat must flow spontaneously "downhill"; that is, from a system at high temperature to one at a lower temperature. This can be expressed scientifically in a number of ways; all the variations of the statement of the second law of thermodynamics. The practical impact of these statements is that energy as it is transformed again and again and transferred from system to system, becomes less and less available for use. Eventually that energy has such low intensity (resides in a medium at such low temperature) that it is no longer available at all to perform a useful function. It has been taken as a general rule of thumb in industrial operations that fluids with temperatures less than 250F are of little value for waste heat extraction; flue gases should not be cooled below 250F (or, better, 300F to provide a safe margin), because of the risk of condensation of corrosive liquids. However, as fuel costs continue to rise, such waste heat can be used economically for space heating and other low temperature applications. It is possible to reverse the direction of spontaneous energy flow by the use of a thermodynamic system known as a heat pump.

This device consists of two heat exchangers, a compressor and an expansion device. A liquid or a mixture of liquid and vapor of a pure chemical species flows through an evaporator, where it absorbs heat at low temperature and in doing so is completely vaporized. The low temperature vapor is compressed by a compressor which requires external work. The work done on the vapor raises its pressure and temperature to a level where its energy becomes available for use. The vapor flows through a condenser where it gives up its energy as it condenses to a liquid. The liquid is then expanded through a device back to the evaporator where the cycle repeats. The heat pump was developed as a space heating system where low temperature energy from the ambient air, water, or earth is raised to heating system temperatures by doing compression work with an electric motor-driven compressor. The performance of the heat pump is ordinarily described in terms of the coefficient of performance or COP, which is defined as:

$$\text{COP} = \frac{\text{Heat transferred in condenser}}{\text{Compressor work}} \qquad 1\text{-}8$$

which in an ideal heat pump is found as:

$$\text{COP} = \frac{T_H}{T_H - T_L} \qquad 1\text{-}9$$

where T_L is the temperature at which waste heat is extracted from the low temperature medium and T_H is the high temperature at which heat is given up by the pump as useful energy. The coefficient of performance expresses the economy of heat transfer.

In the past, the heat pump has not been applied generally to industrial applications. However, several manufacturers are now redeveloping their domestic heat pump systems as well as new equipment for industrial use. The best applications for the device in this new context are not yet clear, but it may well make possible the use of large quantities of low-grade waste heat with relatively small expenditures of work.

Table 1-4. Operation and Application Characteristics of Industrial Heat Exchangers

COMMERCIAL HEAT TRANSFER EQUIPMENT	Low Temperature Sub-Zero – 250°F	Intermediate Temp. 250°F – 1200°F	High Temperature 1200°F – 2000°F	Recovers Moisture	Large Temperature Differentials Permitted	Packaged Units Available	Can Be Retrofit	No Cross-Contamination	Compact Size	Gas-to-Gas Heat Exchange	Gas-to-Liquid Heat Exchanger	Liquid-to-Liquid Heat Exchanger	Corrosive Gases Permitted with Special Construction
Radiation Recuperator			•		•	1	•	•		•			•
Convection Recuperator		•	•		•	•	•	•		•			•
Metallic Heat Wheel	•	•		2		•	•	3	•	•			•
Hygroscopic Heat Wheel	•			•		•	•	3	•	•			
Ceramic Heat Wheel		•	•		•	•	•		•	•			•
Passive Regenerator	•	•			•	•	•	•		•			•
Finned-Tube Heat Exchanger	•	•			•	•	•	•	•		•		4
Tube Shell-and-Tube Exchanger	•	•			•	•	•	•			•	•	
Waste Heat Boilers	•	•			•	•	•				•		4
Heat Pipes	•	•			5	•	•	•	•	•			•

1. Off-the-shelf items available in small capacities only.
2. Controversial subject. Some authorities claim moisture recovery. Do not advise depending on it.
3. With a purge section added, cross-contamination can be limited to less than 1% by mass.
4. Can be constructed of corrosion-resistant materials, but consider possible extensive damage to equipment caused by leaks or tube ruptures.
5. Allowable temperatures and temperature differential limited by the phase equilibrium properties of the internal fluid.

Summary

Table 1-4 presents the collation of a number of significant attributes of the most common types of industrial heat exchangers in matrix form. This matrix allows rapid comparisons to be made in selecting competing types of heat exchangers. The characteristics given in the table for each type of heat exchanger are: allowable temperature range, ability to transfer moisture, ability to withstand large temperature differentials, availability as packaged units, suitability for retrofitting, and compactness and the allowable combinations of heat transfer fluids.

In regard to moisture recovery, it should be emphasized that many of the heat exchangers operating in the low temperature range may condense vapors from the cooled gas stream. Provisions must be made to remove those liquid condensates from the heat exchanger.

REFERENCE

1.   *Waste Heat Management Guidebook*, NBS Handbook 121, U.S. Government Printing Office, Washington, DC 20402.

2

Waste Heat Recovery
Economics

The motivation for companies to invest in waste heat recovery is that they expect the resulting benefits to exceed investment costs. Factors that have recently made such investments attractive are rising fuel costs and curtailment of regular fuel sources which threaten production cutbacks and changeover to other energy sources. In addition, mandatory pollution controls and rising labor costs cut into profits and cause firms to look more closely for ways to control costs.

The kinds of potential benefits which may result from waste heat recovery are listed in Figure 2-1. These benefits were suggested by a preliminary look at existing applications; however, only one, fuel savings, was found in every case examined. The other benefits, savings in capital and maintenance costs on existing equipment, pollution abatement, labor savings, product improvement, and revenue from sales of recovered heat, appear limited to certain applications.

Fuel savings result when waste heat is recovered and used in substitution for newly generated heat or energy. For example, heat from stack flue gas may be recovered by an economizer and used to preheat the input water, thereby reducing the amount of fuel needed for steam generation.

Savings in capital costs for certain items of existing equipment (i.e., regular equipment apart from that required for waste heat

---

Fuel savings

Reduced size, hence lower capital cost, of heating/cooling equipment

Reduced maintenance costs for existing equipment

Reduced costs of production labor

Pollution abatement

Improved product

Revenue from sales of recovered heat or energy

Note: Not all of these benefits will necessarily result from investment in waste heat recovery; in fact, fuel savings may be the only benefit in many applications. However, examples of the other kinds of benefits shown were found in existing applications.

Figure 2-1. Possible Benefits from Waste Heat Recovery

recovery) may be possible if recovered heat reduces the required capacity of the furnace or other heating/cooling equipment.* For example, installation of rooftop thermal recovery equipment on buildings with high ventilation requirements can enable significant reductions in the size, and cost, of the building's heating and cooling system. This potential for savings is probably limited to new plant installations, and does not appear to have received much consideration in industrial applications of heat exchangers.

Reduced maintenance and repair on certain items of existing equipment may, in some instances, be a further benefit of investment in waste heat recovery. The principal impact on the maintenance of existing equipment is likely to result from the planning, engineering, and installation phases of investment in waste heat recovery, when the existing equipment and plant processes are often scrutinized. Existing faults may be identified and corrected; and improved maintenance practices may be extended to existing equipment. While these same effects could be achieved independently of waste heat recovery by a separate inspection of the existing equipment, planning for waste heat recovery provides a catalyst for the inspection. Furthermore, cost of the informational gain is probably

---

*On the other hand, use of heat recovery equipment may increase capital costs of regular equipment by imposing higher temperature loads on it. This effect of increasing costs is included in the listing of costs.

significantly reduced when inspection is performed jointly with the planning for waste heat recovery. Additional effects from waste heat recovery which may reduce maintenance costs include a lowering of the temperature of stack gases.

Another kind of benefit which may result from investment in waste heat recovery is savings in labor costs. Labor savings can result, for example, from a lowering of furnace changeover time (i.e., the time needed to alter furnace temperatures required for a change in production use) by preheating combustion air with waste heat. Savings may also result from faster furnace start-ups, accomplished by similar means. By reducing the amount of labor "downtime," unit labor costs are reduced. (A tradeoff may exist between idling the furnace at higher temperatures during off-duty hours and incurring labor "downtime" during furnace start-ups. If the existing practice is to idle the furnace at high temperatures in order to avoid "downtime," the savings from using waste heat recovery to preheat air would be in terms of fuel reductions rather than lower labor costs.)

Pollution abatement is a beneficial side effect which may result from recovery of waste heat. For example, the pollution abatement process in textile plants will often be facilitated by waste heat recovery. Pollutants (plasticizers) are usually collected by circulating air from the ovens (where fabrics are coated or backed with other materials) through electrostatic precipitators. The air must, however, be cooled to accomplish collection of pollutants. Recovery of waste heat from the air leaving the furnace prior to its entering the precipitators will, consequently, not only provide heat or energy which can be used elsewhere to reduce fuel costs, but will also aid in the collection of pollutants. If it were not for heat recovery, it would be necessary to cool the air by other means, which would generally entail additional fuel consumption. Thus, there is a twofold impact on fuel use from this application of heat recovery.* Another instance

---

*Here we consider only the benefits in fuel savings resulting from pollution abatement effects of waste heat recovery, and not the benefits to the surrounding area from cleaner air emanating from the plant. The emphasis is on private benefits and costs, i.e., those accruing directly to the firm, because private decision makers have traditionally not taken into account all social benefits and costs associated with their investment decisions, i.e., those benefits and costs that accrue to society at large. With the advent of environmental impact statements, however, pollution abatement benefits have become more important to private decision makers.

of pollution abatement as a side effect of waste heat recovery occurs if pollutants are reduced by the higher furnace temperatures resulting from preheating combustion air with waste heat.

The pollution abatement side effects represented by the two preceding examples are distinguishable from the use of systems to recover heat from a pollution abatement process, where recovery of heat does not in itself contribute to pollution abatement. For example, the recovery of waste heat from the incineration of polluting fumes is a method of reducing the cost of pollution abatement by producing a useful by-product from the abatement process. However, the waste heat recovery does not itself contribute to the pollution abatement process and therefore does not yield multiple benefits; the only benefit is the value of the fuel savings from using the recovered heat in other processes.

Product improvement is a further potential side effect of waste heat recovery. For example, by achieving a more stable furnace temperature and a reduction in furnace aeration, use of a recuperator to preheat combustion air may reduce the undesirable scaling of metal products. In absence of preheating combustion air, it would be necessary to invest in improvements to furnace controls or in some other means of preventing scaling; to secure the same product quality.

A final potential benefit from waste heat recovery, as suggested by existing applications, is the generation of revenue from sales of recovered waste heat or energy. In some cases, the recoverable waste heat cannot all be used by the plant itself. Recovery may still be advantageous if there are adjoining plants which are willing to purchase the recovered heat. In this case, the potential benefits are revenue-generating, rather than cost-reducing, and would be measurable in terms of dollars of revenue received.

With the following information—records of past operating levels and expenses, the efficiency of the proposed heat recovery equipment, the level of expected furnace operation, the demand for recycled heat, and the expected price of fuel—it should be possible to predict fairly closely the savings in fuel costs that would result from substituting waste heat for newly-generated heat or energy. Certain of the other potential benefits, such as labor cost savings and product improvement, might be more difficult to estimate.

To evaluate the desirability of an investment, measures of costs are needed to compare with the benefits. Figure 2-2 shows the type of costs which may arise in connection with waste heat recovery.

As may be seen, costs may begin before the waste heat recovery system is installed and extend throughout the period of continued plant operation. In most cases, the major cost item is likely to be the acquisition and installation of the heat exchanger, and should be relatively easy to estimate.

It is important that only those costs and benefits which are attributable to an investment be included in the analysis of that investment. For example, if a plant is required by mandate to add a pollution control apparatus, the decision to add a waste heat recovery system to the pollution control system should not be influenced by the costs of the pollution control system. As a further example, costs of equipment replacement or repair not necessitated by the addition of the waste heat recovery system should not be incorporated into the waste heat evaluation, although it may be undertaken jointly for convenience.

PARTIAL METHODS OF EVALUATION

The simplest procedures which are used by firms to try to evaluate alternative kinds and amounts of investments are visual inspection* and payback period, which are termed "partial" here because they do not fully assess the economic desirability of alternatives. These partial methods may be contrasted with the more complete techniques, discussed in the following section, which take into account factors such as timing of cash flows, risk, and taxation effects —factors which are required for full economic assessment of investments.

---

*There are some investments whose desirability is apparent merely by inspection, and which do not require further economic analysis. An example is an investment characterized by negligible or low costs and a highly certain return. But actions which require significant initial investment and yield benefits over time—as recovery of waste heat is typically characterized—usually require more extensive analysis than visual inspection.

Type of Costs	Examples of Costs
1. Pre-engineering and planning costs	Engineering consultant's fee; in-house manpower and materials to determine type, size, and location of heat exchanger.
2. Acquisition costs of heat recovery equipment	Purchase and installation costs of a recuperator.
3. Acquisition costs of necessary additions to existing equipment	Purchase and installation costs of new controls, burners, stack dampers, and fans to protect the furnace and recuperator from higher temperatures entering the furnace due to preheating of combustion air.
4. Replacement costs	Cost of replacing the inner shell of the recuperator in N years, net of the salvage value of the existing shell.
5. Costs of modification and repair of existing equipment	Cost of repairing furnace doors to overcome greater heat loss resulting from increased pressure due to preheating of combustion air.
6. Space costs	Cost of useful floor space occupied by waste heat steam generator; cost of useful overhead space occupied by evaporator.
7. Costs of production downtime during installation	Loss of output for a week, net of the associated savings in operating costs.
8. Costs of adjustments (debugging)	Lower production; labor costs of debugging.
9. Maintenance costs of new equipment	Costs of servicing the heat exchanger.
10. Property and/or equipment taxes of heat recovery equipment	Additional property tax incurred on capitalized value of recuperator.
11. Change in insurance or hazards costs	Higher insurance rates due to greater fire risks; increased cost of accidents due to more hot spots within a tighter space.

In addition to the above, attention should be given to the length of intended use, expected lives of related equipment, and the flexibility of alternative equipment to future modification and expansion.

Figure 2-2. Potential Costs to Consider in Investing in Waste Heat Recovery

Despite their shortcomings, the partial techniques of analysis may serve a useful purpose. They can provide a first level measure of profitability which is, relatively speaking, quick, simple, and inexpensive to calculate. They may therefore be useful as initial screening devices for eliminating the more obviously uneconomical investments. These partial techniques (particularly the payback method) may also provide needed information concerning certain sensitive features of an investment. But where partial methods are used, the more comprehensive techniques may also be needed to verify the outcome of the evaluations, and to rank alternative projects as to their relative efficiency.

Following are descriptions, examples, and limitations of the payback method and the return on investment method, two of the more popular partial methods.

Payback Method

The payback (also known as the payout or payoff) method determines the number of years required for the invested capital to be offset by resulting benefits. The required number of years is termed the payback, recovery, or break-even period.

The measure is popularly calculated on a before-tax basis and without discounting, i.e., neglecting the opportunity cost of capital. Investment costs are usually defined as first costs, often neglecting salvage value. Benefits are usually defined as the resulting net change in incoming cash flow, or, in the case of a cost-reducing investment like waste heat recovery, as the reduction in net outgoing cash flow.

The payback period is usually calculated as follows:

$$\text{Payback Period (PP)} = \frac{\text{First Cost}}{\text{Yearly Benefits} - \text{Yearly Costs}} \qquad 2\text{-}1$$

SIM 2-1

Calculate the payback period for a furnace recuperator which costs $20,000 to purchase and install, $600/yr on average to operate and maintain, and which is expected to save by preheating combustion air an average of 2000 Mft^3 of burner gas per year at $2.1/Mft^3 (i.e., $4200/yr), as follows:

Answer

$$PP = \frac{20,000}{4200 - 600} = 5.5 \text{ yr}$$

In short, the payback method gives attention to only one attribute of an investment, i.e., the number of years to recover costs, and, as often calculated, does not even provide an accurate measure of this. It is a measure which many firms appear to overemphasize, tending toward shorter and shorter payback requirements. Firms' preference for very short payback to enable them to reinvest in other investment opportunities may in fact lead to a succession of less efficient, short-lived projects.

Despite its limitations, the payback period has *advantages* in that it may provide useful information for evaluating an investment. There are several situations in which the payback method might be particularly appropriate:

(a) A rapid payback may be a prime criterion for judging an investment when financial resources are available to the investor for only a short period of time.

(b) The speculative investor who has a very limited time horizon will usually desire rapid recovery of the initial investment.

(c) Where the expected life of the assets is highly uncertain, determination of the break-even life, i.e., payback period, is helpful in assessing the likelihood of achieving a successful investment.

COMPREHENSIVE METHODS FOR EVALUATING INVESTMENT ALTERNATIVES

There are additional methods of financial analysis which avoid the problems of the partial methods by taking into account total costs and benefits over the life of the investment and the timing of cash flows by discounting.

Discounting of Costs

Investment in waste heat recovery systems, like many capital investments, will generally require a number of expenditures spread over a period of time and will result in cost savings (or revenue receipts) also spread over time. To evaluate correctly the profitability of such investments, it is necessary to convert the various expenditures and receipts to a common basis, because dollars spent or received at different times are not of equal value.

To justify a waste heat recovery expenditure, a knowledge of life-cycle costing is required. The life-cycle cost analysis evaluates the total owning and operating cost. It takes into account the "time value" of money and can incorporate fuel cost escalation into the economic model. This approach is also used to evaluate competitive projects. In other words, the life-cycle cost analysis considers the cost over the life of the system rather than just the first cost.

The Time Value of Money Concept

To compare energy utilization alternatives, it is necessary to convert all cash flow for each measure to an equivalent base. The life-cycle cost analysis takes into account the "time value" of money, thus a dollar in hand today is more valuable than one received at some time in the future. This is why a time value must be placed on all cash flows into and out of the company.

To convert cash from one time to another, any one of the six standard interest factors can be used, as illustrated in Figure 2-3. The seventh factor in the table takes into account escalation. Each factor is described in detail in Figures 2-4 through 2-10.

Interest factors are seldom calculated. They can be determined from computer programs and interest tables (Tables 2-1 through 2-4.) Each factor is defined when the number of periods (n) and interest rate (i) are specified. In the case of the Gradient Present Worth Factor the escalation rate must also be stated.

The three most commonly used methods in life-cycle costing are the annual cost, present worth, and rate-of-return analysis.

In the present worth method a minimum rate of return (i) is stipulated. All future expenditures are converted to present values using the interest factors. The alternative with lowest effective first cost is the most desirable.

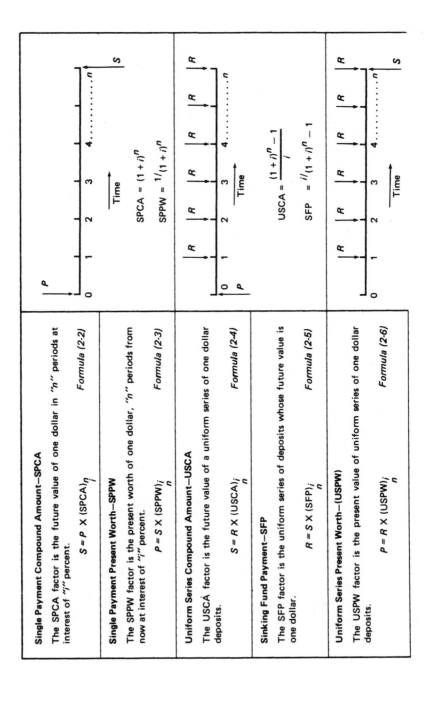

Single Payment Compound Amount—SPCA

The SPCA factor is the future value of one dollar in "n" periods at interest of "i" percent.

$$S = P \times (SPCA)_i^n \qquad \text{Formula (2-2)}$$

$$SPCA = (1 + i)^n$$

Single Payment Present Worth—SPPW

The SPPW factor is the present worth of one dollar, "n" periods from now at interest of "i" percent.

$$P = S \times (SPPW)_i^n \qquad \text{Formula (2-3)}$$

$$SPPW = 1/(1 + i)^n$$

Uniform Series Compound Amount—USCA

The USCA factor is the future value of a uniform series of one dollar deposits.

$$S = R \times (USCA)_i^n \qquad \text{Formula (2-4)}$$

$$USCA = \frac{(1 + i)^n - 1}{i}$$

Sinking Fund Payment—SFP

The SFP factor is the uniform series of deposits whose future value is one dollar.

$$R = S \times (SFP)_i^n \qquad \text{Formula (2-5)}$$

$$SFP = i/(1 + i)^n - 1$$

Uniform Series Present Worth—(USPW)

The USPW factor is the present value of uniform series of one dollar deposits.

$$P = R \times (USPW)_i^n \qquad \text{Formula (2-6)}$$

Capital Recovery—CR

The CR factor is the uniform series of deposits whose present value is one dollar.

$$R = P \times (CR)_{i,n} \qquad \text{Formula (2-7)}$$

Gradient Present Worth—GPW

The GPW factor is the present value of a gradient series.

$$P = R \times (GPW)_{i,n} \qquad \text{Formula (2-8)}$$

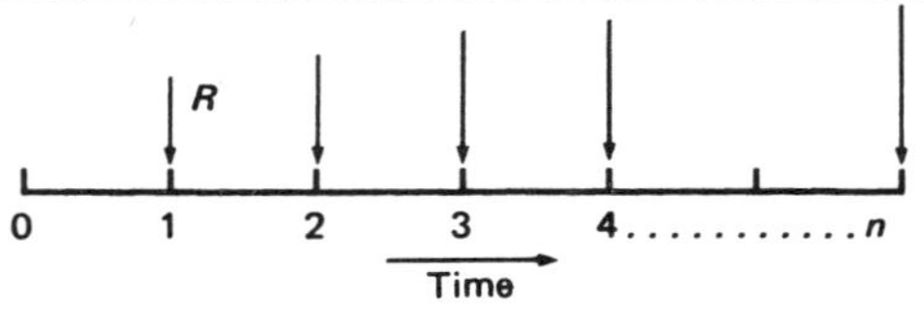

$$USPW = \frac{(1+i)^n - 1}{i(1+i)^n}$$

$$CR = \frac{i(1+i)^n}{(1+i)^n - 1}$$

$$GPW = \frac{1+e}{i-e}\left[1 - \left(\frac{1+e}{1+i}\right)^n\right]$$

NOTES

where

P is the present worth (occurs at the beginning of the interest period).

S is the future worth (occurs at the end).

n is the number of periods that the interest is compounded.

i is the interest rate or desired rate of return.

R is the uniform series of deposits (occurs at the end of the interest period.)

e is the escalation rate.

Figure 2-3. Interest Factors

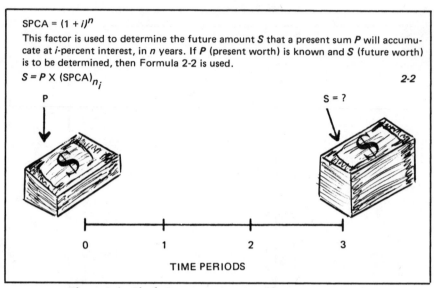

$SPCA = (1 + i)^n$

This factor is used to determine the future amount S that a present sum P will accumucate at i-percent interest, in n years. If P (present worth) is known and S (future worth) is to be determined, then Formula 2-2 is used.

$$S = P \times (SPCA)_{n_i} \qquad\qquad 2\text{-}2$$

P

S = ?

0   1   2   3

TIME PERIODS

Figure 2-4. Single Payment Compound Amount (SPCA)

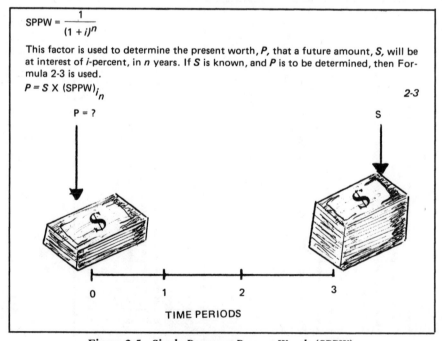

$$SPPW = \frac{1}{(1 + i)^n}$$

This factor is used to determine the present worth, P, that a future amount, S, will be at interest of i-percent, in n years. If S is known, and P is to be determined, then Formula 2-3 is used.

$$P = S \times (SPPW)_{i_n} \qquad\qquad 2\text{-}3$$

P = ?

S

0   1   2   3

TIME PERIODS

Figure 2-5. Single Payment Present Worth (SPPW)

$$USCA = \frac{(1 + i)^n - 1}{i}$$

This factor is used to determine the amount S that an equal annual payment R will accumulate to in n years at i-percent interest. If R (uniform annual payment) is known, S (the future worth of these payments), is required, then Formula 2-4 is used.

$S = R \times (USCA)_{i_n}$ 2-4

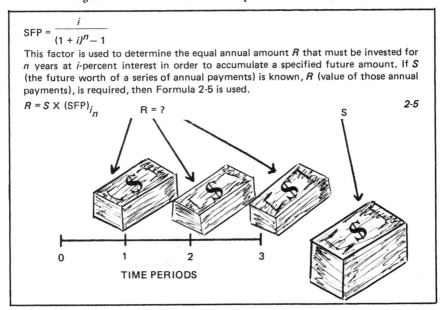

Figure 2-6. Uniform Series Compound Amount (USCA)

$$SFP = \frac{i}{(1 + i)^n - 1}$$

This factor is used to determine the equal annual amount R that must be invested for n years at i-percent interest in order to accumulate a specified future amount. If S (the future worth of a series of annual payments) is known, R (value of those annual payments), is required, then Formula 2-5 is used.

$R = S \times (SFP)_{i_n}$ 2-5

Figure 2-7. Sinking Fund Payment (SFP)

$$USPW = \frac{(1 + i)^n - 1}{i(1 + i)^n}$$

This factor is used to determine the present amount P that can be paid by equal payments of R (uniform annual payment) at i-percent interest, for n years. If R is known P is required, then Formula 2-6 is used.

$$P = R \times (USPW)_{i_n} \qquad\qquad 2\text{-}6$$

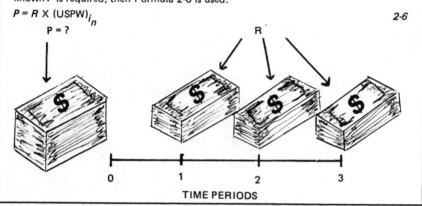

Figure 2-8. Uniform Series Present Worth (USPW)

$$CR = \frac{i(1 + i)^n}{(1 + i)^n - 1}$$

This factor is used to determine an annual payment R required to pay off a present amount P at i-percent interest, for n years. If the present sum of money, P spent today, and uniform payment R needed to pay back P over a stated period of time is required, then Formula 2-7 is used.

$$R = P \times (CR)_{i_n} \qquad\qquad 2\text{-}7$$

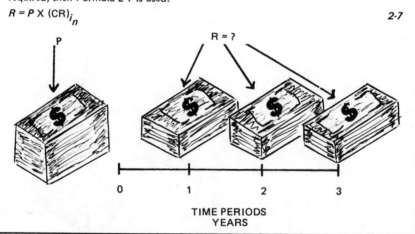

Figure 2-9. Capital Recovery (CR)

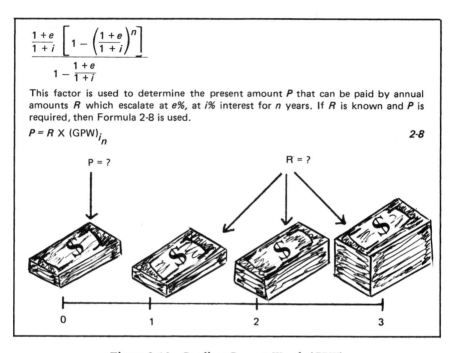

$$\frac{\dfrac{1+e}{1+i}\left[1-\left(\dfrac{1+e}{1+i}\right)^{n}\right]}{1-\dfrac{1+e}{1+i}}$$

This factor is used to determine the present amount P that can be paid by annual amounts R which escalate at $e\%$, at $i\%$ interest for n years. If R is known and P is required, then Formula 2-8 is used.

$P = R \times (GPW)_{i_n}$ 2-8

$P = ?$

$R = ?$

0	1	2	3

Figure 2-10. Gradient Present Worth (GPW)

A similar procedure is implemented in the annual cost method. The difference is that the first cost is converted to an annual expenditure. The alternative with lowest effective annual cost is the most desirable.

In the rate-of-return method, a trial-and-error procedure is usually required. Interpolation from the interest tables can determine what rate of return (i) will give an interest factor which will make the overall cash flow balance. The rate-of-return analysis gives a good indication of the overall ranking of independent alternates.

The effect of escalation in fuel costs can influence greatly the final decision. When an annual cost grows at a steady rate it may be treated as a gradient and the Gradient Present Worth Factor can be used.

Special appreciation is given to Rudolph R. Yaneck and Dr. Robert Brown for use of their specially designed interest and escalation tables used in this text.

Table 2-1. 12% Interest Factor

Period n	Single-payment compound-amount (SPCA) Future value of $1 $(1 + i)^n$	Single-payment present-worth (SPPW) Present value of $1 $\dfrac{1}{(1 + i)^n}$	Uniform-series compound-amount (USCA) Future value of uniform series of $1 $\dfrac{(1 + i)^n - 1}{i}$	Sinking-fund payment (SFP) Uniform series whose future value is $1 $\dfrac{i}{(1 + i)^n - 1}$	Capital recovery (CR) Uniform series with present value of $1 $\dfrac{i(1 + i)^n}{(1 + i)^n - 1}$	Uniform-series present-worth (USPW) Present value of uniform series of $1 $\dfrac{(1 + i)^n - 1}{i(1 + i)^n}$
1	1.120	0.8929	1.000	1.00000	1.12000	0.893
2	1.254	0.7972	2.120	0.47170	0.59170	1.690
3	1.405	0.7118	3.374	0.29635	0.41635	2.402
4	1.574	0.6355	4.779	0.20923	0.32923	3.037
5	1.762	0.5674	6.353	0.15741	0.27741	3.605
6	1.974	0.5066	8.115	0.12323	0.24323	4.111
7	2.211	0.4523	10.089	0.09912	0.21912	4.564
8	2.476	0.4039	12.300	0.08130	0.20130	4.968
9	2.773	0.3606	14.776	0.06768	0.18768	5.328
10	3.106	0.3220	17.549	0.05698	0.17698	5.650
11	3.479	0.2875	20.655	0.04842	0.16842	5.938
12	3.896	0.2567	24.133	0.04144	0.16144	6.194
13	4.363	0.2292	28.029	0.03568	0.15568	6.424
14	4.887	0.2046	32.393	0.03087	0.15087	6.628
15	5.474	0.1827	37.280	0.02682	0.14682	6.811
16	6.130	0.1631	42.753	0.02339	0.14339	6.974
17	6.866	0.1456	48.884	0.02046	0.14046	7.120
18	7.690	0.1300	55.750	0.01794	0.13794	7.250
19	8.613	0.1161	63.440	0.01576	0.13576	7.366
20	9.646	0.1037	72.052	0.01388	0.13388	7.469
21	10.804	0.0926	81.699	0.01224	0.13224	7.562
22	12.100	0.0826	92.503	0.01081	0.13081	7.645
23	13.552	0.0738	104.603	0.00956	0.12956	7.718
24	15.179	0.0659	118.155	0.00846	0.12846	7.784
25	17.000	0.0588	133.334	0.00750	0.12750	7.843
26	19.040	0.0525	150.334	0.00665	0.12665	7.896
27	21.325	0.0469	169.374	0.00590	0.12590	7.943
28	23.884	0.0419	190.699	0.00524	0.12524	7.984
29	26.750	0.0374	214.583	0.00466	0.12466	8.022
30	29.960	0.0334	241.333	0.00414	0.12414	8.055
35	52.800	0.0189	431.663	0.00232	0.12232	8.176
40	93.051	0.0107	767.091	0.00130	0.12130	8.244
45	163.988	0.0061	1358.230	0.00074	0.12074	8.283
50	289.002	0.0035	2400.018	0.00042	0.12042	8.304
55	509.321	0.0020	4236.005	0.00024	0.12024	8.317
60	897.597	0.0011	7471.641	0.00013	0.12013	8.324
65	1581.872	0.0006	13173.937	0.00008	0.12008	8.328
70	2787.800	0.0004	23223.332	0.00004	0.12004	8.330
75	4913.056	0.0002	40933.799	0.00002	0.12002	8.332
80	8658.483	0.0001	72145.692	0.00001	0.12001	8.332

Table 2-2. 15% Interest Factor

Period n	Single-payment compound-amount (SPCA)	Single-payment present-worth (SPPW)	Uniform-series compound-amount (USCA)	Sinking-fund payment (SFP)	Capital recovery (CR)	Uniform-series present-worth (USPW)
	Future value of $1 $(1 + i)^n$	Present value of $1 $\dfrac{1}{(1+i)^n}$	Future value of uniform series of $1 $\dfrac{(1+i)^n - 1}{i}$	Uniform series whose future value is $1 $\dfrac{i}{(1+i)^n - 1}$	Uniform series with present value of $1 $\dfrac{i(1+i)^n}{(1+i)^n - 1}$	Present value of uniform series of $1 $\dfrac{(1+i)^n - 1}{i(1+i)^n}$
1	1.150	0.8696	1.000	1.00000	1.15000	0.870
2	1.322	0.7561	2.150	0.46512	0.61512	1.626
3	1.521	0.6575	3.472	0.28798	0.43798	2.283
4	1.749	0.5718	4.993	0.20027	0.35027	2.855
5	2.011	0.4972	6.742	0.14832	0.29832	3.352
6	2.313	0.4323	8.754	0.11424	0.26424	3.784
7	2.660	0.3759	11.067	0.09036	0.24036	4.160
8	3.059	0.3269	13.727	0.07285	0.22285	4.487
9	3.518	0.2843	16.786	0.05957	0.20957	4.772
10	4.046	0.2472	20.304	0.04925	0.19925	5.019
11	4.652	0.2149	24.349	0.04107	0.19107	5.234
12	5.350	0.1869	29.002	0.03448	0.18448	5.421
13	6.153	0.1625	34.352	0.02911	0.17911	5.583
14	7.076	0.1413	40.505	0.02469	0.17469	5.724
15	8.137	0.1229	47.580	0.02102	0.17102	5.847
16	9.358	0.1069	55.717	0.01795	0.16795	5.954
17	10.761	0.0929	65.075	0.01537	0.16537	6.047
18	12.375	0.0808	75.836	0.01319	0.16319	6.128
19	14.232	0.0703	88.212	0.01134	0.16134	6.198
20	16.367	0.0611	102.444	0.00976	0.15976	6.259
21	18.822	0.0531	118.810	0.00842	0.15842	6.312
22	21.645	0.0462	137.632	0.00727	0.15727	6.359
23	24.891	0.0402	159.276	0.00628	0.15628	6.399
24	28.625	0.0349	184.168	0.00543	0.15543	6.434
25	32.919	0.0304	212.793	0.00470	0.15470	6.464
26	37.857	0.0264	245.712	0.00407	0.15407	6.491
27	43.535	0.0230	283.569	0.00353	0.15353	6.514
28	50.066	0.0200	327.104	0.00306	0.15306	6.534
29	57.575	0.0174	377.170	0.00265	0.15265	6.551
30	66.212	0.0151	434.745	0.00230	0.15230	6.566
35	133.176	0.0075	881.170	0.00113	0.15113	6.617
40	267.864	0.0037	1779.090	0.00056	0.15056	6.642
45	538.769	0.0019	3585.128	0.00028	0.15028	6.654
50	1083.657	0.0009	7217.716	0.00014	0.15014	6.661
55	2179.622	0.0005	14524.148	0.00007	0.15007	6.664
60	4383.999	0.0002	29219.992	0.00003	0.15003	6.665
65	8817.787	0.0001	58778.583	0.00002	0.15002	6.666

Table 2-3. 20% Interest Factor

Period n	Single-payment compound-amount (SPCA)	Single-payment present-worth (SPPW)	Uniform-series compound-amount (USCA)	Sinking-fund payment (SFP)	Capital recovery (CR)	Uniform-series present-worth (USPW)
	Future value of $1 $(1 + i)^n$	Present value of $1 $\dfrac{1}{(1 + i)^n}$	Future value of uniform series of $1 $\dfrac{(1 + i)^n - 1}{i}$	Uniform series whose future value is $1 $\dfrac{i}{(1 + i)^n - 1}$	Uniform series with present value of $1 $\dfrac{i(1 + i)^n}{(1 + i)^n - 1}$	Present value of uniform series of $1 $\dfrac{(1 + i)^n - 1}{i(1 + i)^n}$
1	1.200	0.8333	1.000	1.00000	1.20000	0.833
2	1.440	0.6944	2.200	0.45455	0.65455	1.528
3	1.728	0.5787	3.640	0.27473	0.47473	2.106
4	2.074	0.4823	5.368	0.18629	0.38629	2.589
5	2.488	0.4019	7.442	0.13438	0.33438	2.991
6	2.986	0.3349	9.930	0.10071	0.30071	3.326
7	3.583	0.2791	12.916	0.07742	0.27742	3.605
8	4.300	0.2326	16.499	0.06061	0.26061	3.837
9	5.160	0.1938	20.799	0.04808	0.24808	4.031
10	6.192	0.1615	25.959	0.03852	0.23852	4.192
11	7.430	0.1346	32.150	0.03110	0.23110	4.327
12	8.916	0.1122	39.581	0.02526	0.22526	4.439
13	10.699	0.0935	48.497	0.02062	0.22062	4.533
14	12.839	0.0779	59.196	0.01689	0.21689	4.611
15	15.407	0.0649	72.035	0.01388	0.21388	4.675
16	18.488	0.0541	87.442	0.01144	0.21144	4.730
17	22.186	0.0451	105.931	0.00944	0.20944	4.775
18	26.623	0.0376	128.117	0.00781	0.20781	4.812
19	31.948	0.0313	154.740	0.00646	0.20646	4.843
20	38.338	0.0261	186.688	0.00536	0.20536	4.870
21	46.005	0.0217	225.026	0.00444	0.20444	4.891
22	55.206	0.0181	271.031	0.00369	0.20369	4.909
23	66.247	0.0151	326.237	0.00307	0.20307	4.925
24	79.497	0.0126	392.484	0.00255	0.20255	4.937
25	95.396	0.0105	471.981	0.00212	0.20212	4.948
26	114.475	0.0087	567.377	0.00176	0.20176	4.956
27	137.371	0.0073	681.853	0.00147	0.20147	4.964
28	164.845	0.0061	819.223	0.00122	0.20122	4.970
29	197.814	0.0051	984.068	0.00102	0.20102	4.975
30	237.376	0.0042	1181.882	0.00085	0.20085	4.979
35	590.668	0.0017	2948.341	0.00034	0.20034	4.992
40	1469.772	0.0007	7343.858	0.00014	0.20014	4.997
45	3657.262	0.0003	18281.310	0.00005	0.20005	4.999
50	9100.438	0.0001	45497.191	0.00002	0.20002	4.999

Table 2-4. 25% Interest Factor

Period n	Single-payment compound-amount (SPCA) Future value of $1 $(1 + i)^n$	Single-payment present-worth (SPPW) Present value of $1 $\dfrac{1}{(1 + i)^n}$	Uniform-series compound amount (USCA) Future value of uniform series of $1 $\dfrac{(1 + i)^n - 1}{i}$	Sinking-fund payment (SFP) Uniform series whose future value is $1 $\dfrac{i}{(1 + i)^n - 1}$	Capital recovery (CR) Uniform series with present value of $1 $\dfrac{i(1 + i)^n}{(1 + i)^n - 1}$	Uniform-series present-worth (USPW) Present value of uniform series of $1 $\dfrac{(1 + i)^n - 1}{i(1 + i)^n}$
1	1.250	0.8000	1.000	1.00000	1.25000	0.800
2	1.562	0.6400	2.250	0.44444	0.69444	1.440
3	1.953	0.5120	3.812	0.26230	0.51230	1.952
4	2.441	0.4096	5.766	0.17344	0.42344	2.362
5	3.052	0.3277	8.207	0.12185	0.37185	2.689
6	3.815	0.2621	11.259	0.08882	0.33882	2.951
7	4.768	0.2097	15.073	0.06634	0.31634	3.161
8	5.960	0.1678	19.842	0.05040	0.30040	3.329
9	7.451	0.1342	25.802	0.03876	0.28876	3.463
10	9.313	0.1074	33.253	0.03007	0.28007	3.571
11	11.642	0.0859	42.566	0.02349	0.27349	3.656
12	14.552	0.0687	54.208	0.01845	0.26845	3.725
13	18.190	0.0550	68.760	0.01454	0.26454	3.780
14	22.737	0.0440	86.949	0.01150	0.26150	3.824
15	28.422	0.0352	109.687	0.00912	0.25912	3.859
16	35.527	0.0281	138.109	0.00724	0.25724	3.887
17	44.409	0.0225	173.636	0.00576	0.25576	3.910
18	55.511	0.0180	218.045	0.00459	0.25459	3.928
19	69.389	0.0144	273.556	0.00366	0.25366	3.942
20	86.736	0.0115	342.945	0.00292	0.25292	3.954
21	108.420	0.0092	429.681	0.00233	0.25233	3.963
22	135.525	0.0074	538.101	0.00186	0.25186	3.970
23	169.407	0.0059	673.626	0.00148	0.25148	3.976
24	211.758	0.0047	843.033	0.00119	0.25119	3.981
25	264.698	0.0038	1054.791	0.00095	0.25095	3.985
26	330.872	0.0030	1319.489	0.00076	0.25076	3.988
27	413.590	0.0024	1650.361	0.00061	0.25061	3.990
28	516.988	0.0019	2063.952	0.00048	0.25048	3.992
29	646.235	0.0015	2580.939	0.00039	0.25039	3.994
30	807.794	0.0012	3227.174	0.00031	0.25031	3.995
35	2465.190	0.0004	9856.761	0.00010	0.25010	3.998
40	7523.164	0.0001	30088.655	0.00003	0.25003	3.999

Table 2-5. 30% Interest Factor

Period n	Single-payment compound-amount (SPCA)	Single-payment present-worth (SPPW)	Uniform-series compound-amount (USCA)	Sinking-fund payment (SFP)	Capital recovery (CR)	Uniform-series present-worth (USPW)
	Future value of $1 $(1+i)^n$	Present value of $1 $\dfrac{1}{(1+i)^n}$	Future value of uniform series of $1 $\dfrac{(1+i)^n-1}{i}$	Uniform series whose future value is $1 $\dfrac{i}{(1+i)^n-1}$	Uniform series with present value of $1 $\dfrac{i(1+i)^n}{(1+i)^n-1}$	Present value of uniform series of $1 $\dfrac{(1+i)^n-1}{i(1+i)^n}$
1	1.300	0.7692	1.000	1.00000	1.30000	0.769
2	1.690	0.5917	2.300	0.43478	0.73478	1.361
3	2.197	0.4552	3.990	0.25063	0.55063	1.816
4	2.856	0.3501	6.187	0.16163	0.46163	2.166
5	3.713	0.2693	9.043	0.11058	0.41058	2.436
6	4.827	0.2072	12.756	0.07839	0.37839	2.643
7	6.275	0.1594	17.583	0.05687	0.35687	2.802
8	8.157	0.1226	23.858	0.04192	0.34192	2.925
9	10.604	0.0943	32.015	0.03124	0.33124	3.019
10	13.786	0.0725	42.619	0.02346	0.32346	3.092
11	17.922	0.0558	56.405	0.01773	0.31773	3.147
12	23.298	0.0429	74.327	0.01345	0.31345	3.190
13	30.288	0.0330	97.625	0.01024	0.31024	3.223
14	39.374	0.0254	127.913	0.00782	0.30782	3.249
15	51.186	0.0195	167.286	0.00598	0.30598	3.268
16	66.542	0.0150	218.472	0.00458	0.30458	3.283
17	86.504	0.0116	285.014	0.00351	0.30351	3.295
18	112.455	0.0089	371.518	0.00269	0.30269	3.304
19	146.192	0.0068	483.973	0.00207	0.30207	3.311
20	190.050	0.0053	630.165	0.00159	0.30159	3.316
21	247.065	0.0040	820.215	0.00122	0.30122	3.320
22	321.184	0.0031	1067.280	0.00094	0.30094	3.323
23	417.539	0.0024	1388.464	0.00072	0.30072	3.325
24	542.801	0.0018	1806.003	0.00055	0.30055	3.327
25	705.641	0.0014	2348.803	0.00043	0.30043	3.329
26	917.333	0.0011	3054.444	0.00033	0.30033	3.330
27	1192.533	0.0008	3971.778	0.00025	0.30025	3.331
28	1550.293	0.0006	5164.311	0.00019	0.30019	3.331
29	2015.381	0.0005	6714.604	0.00015	0.30015	3.332
30	2619.996	0.0004	8729.985	0.00011	0.30011	3.332
35	9727.860	0.0001	32422.868	0.00003	0.30003	3.333

Table 2-6. 40% Interest Factor

Period n	Single-payment compound-amount (SPCA)	Single-payment present-worth (SPPW)	Uniform-series compound-amount (USCA)	Sinking-fund payment (SFP)	Capital recovery (CR)	Uniform-series present-worth (USPW)
	Future value of $1 $(1 + i)^n$	Present value of $1 $\dfrac{1}{(1 + i)^n}$	Future value of uniform series of $1 $\dfrac{(1 + i)^n - 1}{i}$	Uniform series whose future value is $1 $\dfrac{i}{(1 + i)^n - 1}$	Uniform series with present value of $1 $\dfrac{i(1 + i)^n}{(1 + i)^n - 1}$	Present value of uniform series of $1 $\dfrac{(1 + i)^n - 1}{i(1 + i)^n}$
1	1.400	0.7143	1.000	1.00000	1.40000	0.714
2	1.960	0.5102	2.400	0.41667	0.81667	1.224
3	2.744	0.3644	4.360	0.22936	0.62936	1.589
4	3.842	0.2603	7.104	0.14077	0.54077	1.849
5	5.378	0.1859	10.946	0.09136	0.49136	2.035
6	7.530	0.1328	16.324	0.06126	0.46126	2.168
7	10.541	0.0949	23.853	0.04192	0.44192	2.263
8	14.758	0.0678	34.395	0.02907	0.42907	2.331
9	20.661	0.0484	49.153	0.02034	0.42034	2.379
10	28.925	0.0346	69.814	0.01432	0.41432	2.414
11	40.496	0.0247	98.739	0.01013	0.41013	2.438
12	56.694	0.0176	139.235	0.00718	0.40718	2.456
13	79.371	0.0126	195.929	0.00510	0.40510	2.469
14	111.120	0.0090	275.300	0.00363	0.40363	2.478
15	155.568	0.0064	386.420	0.00259	0.40259	2.484
16	217.795	0.0046	541.988	0.00185	0.40185	2.489
17	304.913	0.0033	759.784	0.00132	0.40132	2.492
18	426.879	0.0023	1064.697	0.00094	0.40094	2.494
19	597.630	0.0017	1491.576	0.00067	0.40067	2.496
20	836.683	0.0012	2089.206	0.00048	0.40048	2.497
21	1171.356	0.0009	2925.889	0.00034	0.40034	2.498
22	1639.898	0.0006	4097.245	0.00024	0.40024	2.498
23	2295.857	0.0004	5737.142	0.00017	0.40017	2.499
24	3214.200	0.0003	8032.999	0.00012	0.40012	2.499
25	4499.880	0.0002	11247.199	0.00009	0 40009	2.499
26	6299.831	0.0002	15747.079	0.00006	0.40006	2.500
27	8819.764	0.0001	22046.910	0.00005	0.40005	2.500

Table 2-7. 20 Year Escalation Table

Source: Brown & Yanuck

Present Worth of a Series of Escalating Payments Compounded Annually
Discount-Escalation Factors for N = 20 Years

Discount Rate	Annual Escalation Rate					
	.10	.12	.14	.16	.18	.20
0.10	20.000000	24.295450	29.722090	36.592170	45.308970	56.383330
0.11	18.213210	22.002090	26.776150	32.799710	40.417480	50.067940
0.12	16.642370	20.000000	24.210030	29.505430	36.181240	44.614710
0.13	15.259850	18.243100	21.964990	26.634490	32.502270	39.891400
0.14	14.038630	16.654830	20.000000	24.127100	29.298170	35.789680
0.15	12.957040	15.329770	18.271200	21.929940	26.498510	32.218060
0.16	11.995640	14.121040	16.746150	20.000000	24.047720	29.098950
0.17	11.138940	13.048560	15.397670	18.300390	21.894660	26.369210
0.18	10.373120	12.053400	14.201180	16.795710	20.000000	23.970940
0.19	9.686791	11.240870	13.137510	15.463070	18.326720	21.860120
0.20	9.069737	10.477430	12.186860	14.279470	16.844020	20.000000

Discount Rate	Annual Escalation Rate					
	.10	.12	.14	.16	.18	.20
0.21	8.513605	9.792256	11.340570	13.224610	15.527270	18.353210
0.22	8.010912	9.175267	10.579620	12.282120	14.355520	16.890730
0.23	7.555427	8.618459	9.895583	11.438060	13.309280	15.589300
0.24	7.141531	8.114476	9.278916	10.679310	12.373300	14.429370
0.25	6.764528	7.657278	8.721467	9.997057	11.533310	13.392180
0.26	6.420316	7.241402	8.216490	9.380883	10.778020	12.462340
0.27	6.105252	6.862203	7.757722	8.823063	10.096710	11.626890
0.28	5.816151	6.515563	7.339966	8.316995	9.480940	10.874120
0.29	5.550301	6.198027	6.958601	7.856833	8.922847	10.194520
0.30	5.305312	5.906440	6.609778	7.437339	8.416060	9.579437
0.31	5.079039	5.638064	6.289875	7.054007	7.954518	9.021190
0.32	4.869585	5.390575	5.995840	6.702967	7.533406	8.513612
0.33	4.675331	5.161809	5.725066	6.380829	7.148199	8.050965
0.34	4.494838	4.949990	5.475180	6.084525	6.795200	7.628322

When life-cycle costing is used to compare several alternatives the differences between costs are important. For example, if one alternate forces additional maintenance or an operating expense to occur, then these factors as well as energy costs need to be included. Remember, what was previously spent for the item to be replaced is irrelevant. The only factor to be considered is whether the new cost can be justified based on projected savings over its useful life.

SIM 2-2

An economizer is being considered to recover heat from the combustion stack. The total installed cost of the economizer is $60,000 with a useful life of 20 years. The fuel cost is reduced from $240,000 to $216,000 with the addition of the economizer. Is the investment justified based on a minimum rate of return before taxes of 20%? Solve this problem using the present worth, annual cost, and rate-of-return methods.

Present Worth Method

		Alternate 1 Present Method	Alternate 2 Economizer
(1)	First Cost	—	$ 60,000
(2)	Annual Cost	$ 240,000	216,000
(3)	USPW i = 20% n = 20 yrs (Table 2-3)	4.870	4.870
(4)	Present Worth of annual costs = (2) × (3)	$ 1.168 × 10^6	$1.051 × 10^6
(5)	Present Worth = (1) + (4)	$ 1.168 × 10^6	$1.111 × 10^6

Choose Alternate with lowest First Cost.

Annual Cost Method

		Alternate 1	Alternate 2
(1)	First Cost	—	$ 60,000
(2)	Annual Cost	$240,000	216,000
(3)	CR	.2	.2
	i = 20%		
	n = 20 yrs		
	(Table 2-3)		
(4)	Annual Worth of		
	First Cost	—	$ 12,000
	(1) X (3)		
(5)	Annual Cost	$240,000	$228,000
	(2) + (4)		

Choose Alternate with lowest annual cost.

Rate-of-Return Method

$$P = R \times USPW$$
$$= (\$240,000 - \$216,000) \times USPW$$

$$USPW = \frac{60,000}{24,000} = 2.5$$

What value i will make USPW = 2.5?

From Table 2-6
i = 40%

SIM 2-3

Show the effect of 10% escalation on the rate of return analysis given the

Heat recovery equipment investment = $20,000
After tax savings = $ 2,600
Equipment life (n) = 20 years

Analysis

Without escalation

$$CR = \frac{R}{P} = \frac{2,600}{20,000} = .13$$

From Table 2-1, the rate of return is 12%.

With 10% escalation assumed:

$$GPW = \frac{P}{G} = \frac{20,000}{2,600} = 7.69$$

From Table 2-7, the rate of return is approximately 23%.

Thus we see that taking into account a modest escalation rate can dramatically affect the justification of the project.

THE EFFECT OF INCOME TAXES

Income taxes can have a profound impact on optimum investment decisions. By changing the effective values to the firm of the revenues and costs associated with an investment, taxes can reverse the relative profitability of alternative projects as evaluated apart from taxes, and they can alter the optimal size of investments. Taxes are, therefore, an important element in the conomic evaluation of investment in waste heat recovery systems.

If, for example, an investment results in increased taxable revenue, the effective value of the additional revenue to the firm is reduced by taxation.

Tax-deductible expenses such as maintenance, energy, operating costs, insurance and property taxes reduce the income subject to taxes.

For the after tax life-cycle cost analysis and payback analysis the actual incurred annual savings is given as follows:

$$AS = (1-I) E + ID \qquad\qquad 2-9$$

Where:

AS = yearly annual after tax savings (excluding effect of tax credit)

E = yearly annual energy savings (difference between original expenses and expenses after modification)

D = annual depreciation rate
I = income tax bracket

Formula 2-9 takes into account that the yearly annual energy savings is partially offset by additional taxes which must be paid due to reduced operating expenses. On the other hand, the depreciation allowance reduces taxes directly.

Depreciation

Depreciation affects the "accounting procedure" for determining profits and losses and the income tax of a company. In other words, for tax purposes the expenditure for an asset such as a pump or motor cannot be fully expensed in its first year. The original investment must be charged off for tax purposes over the useful life of the asset. A company usually wishes to expense an item as quickly as possible.

The Internal Revenue Service allows several methods for determining the annual depreciation rate.

As a result of The Economic Recovery Tax Act of 1981 there is a brand new, generally faster method of writing off the cost of tangible property used in business or held for the production of income. It's called the "Accelerated Cost Recovery System" or ACRS. The new system is generally applicable to eligible property (called "recovery property") placed in service on or after January 1, 1981. So it may apply to depreciable property that you've already purchased.

Recovery property is divided into four classes: 3-year, 5-year, 10-year, and 15-year property. For example:

- 3-year: Cars, light duty trucks and certain other short-lived personal property
- 5-year: Most machinery and equipment
- 15-year: Buildings

For each class there is a standard set of recovery deductions (i.e., depreciation with a new name) to be taken over a fixed recovery period.

Tax Credit

A tax credit encourages capital investment. Essentially the tax credit lowers the income tax paid by the tax credit to an upper limit.

In addition to the investment tax credit, the Business Energy Tax Credit as a result of the National Energy Plan, can also be taken. The Business Energy Tax Credit applies to industrial investment in alternative energy property such as boilers for coal, heat conservation, and recycling equipment. The tax credit substantially increases the investment merit of the investment since it lowers the *bottom* line on the tax form. With rapid changes in tax laws be sure to check the latest revision before applying the tax credit.

After-Tax Analysis

To compute a rate of return which accounts for taxes, depreciation, escalation, and tax credits, a cash-flow analysis is usually required. This method analyzes all transactions including first and operating costs. To determine the after-tax rate of return, a trial and error, or computer analysis, is required.

The Present Worth Factors summarized in Tables 2-1 through 2-6 can be used for this analysis. All money is converted to the present assuming an interest rate. The summation of all present dollars should equal zero when the correct interest rate is selected, as illustrated in Figure 2-11.

SIM 2-4

Comment on the after-tax rate of return for the installation of a heat-recovery system without tax credit given the following:

- First Cost $100,000
- Year Savings   40,000
- Depreciation rate — $20,000 per year
- Income tax bracket — 46%

Year	1 Investment	2 Tax Credit	3 After Tax Savings (AS)	4 Single Payment Present Worth Factor	(2 + 3) X 4 Present Worth
0	−P				−P
1		+TC	AS_1	$SPPW_1$	$+P_1$
2			AS_2	$SPPW_2$	P_2
3			AS_3	$SPPW_3$	P_3
4			AS_4	$SPPW_4$	P_4
Total					ΣP

$$AS = (1 - I) E + ID$$

Trial & Error Solution:
Correct i when $\Sigma P = 0$

Figure 2-11. Cash Flow Rate of Return Analysis

Analysis

$$D = \$20,000$$
$$AS = (1-I) E + ID = .54(40,000) + .46(20,000)$$
$$= 21,600 + 9,200 = 30,800$$

First Trial i = 20%

Investment	After Tax Savings	SPPW 20%	PW
0 − 100,000			−100,000
1	30,800	.833	25,656
2	30,800	.694	21,375
3	30,800	.578	17,802
4	30,800	.482	14,845
5	30,800	.401	12,350
		$\Sigma -$	7,972

Since summation is negative a higher present worth factor is required. Next try is 15%.

Investment	After Tax Savings	SPPW 15%	PW
0 −100,000			−100,000
1	30,800	.869	+ 26,765
2	30,800	.756	+ 23,284
3	30,800	.657	+ 20,235
4	30,800	.571	+ 17,586
5	30,800	.497	+ 15,307
			+ 3,177

Since rate of return is bracketed, linear interpolation will be used.

$$\frac{3177 + 7971}{-5} = \frac{3177 - 0}{15 - i\%}$$

$$i = \frac{3177}{2229.6} + 15 = 16.4\%$$

**Short-Cut Method for
After-Tax Rate of Return**

Using the Cash Flow Rate of Return Analysis, a series of curves is developed in Figure 2-12 to aid in energy economic decision making.

The curves indicate the capital investment which can be justified for each $1,000 saved given a fuel escalation of $e\%$.

SIM 2-5

It is desired to have an after-tax savings of 15%. Comment on the investment which can be justified if it is assumed that the fuel escalation rate should not be considered and the annual energy savings is $4,000 with an equipment economic life of 15 years.

Comment on the above, assuming a 14% fuel escalation.

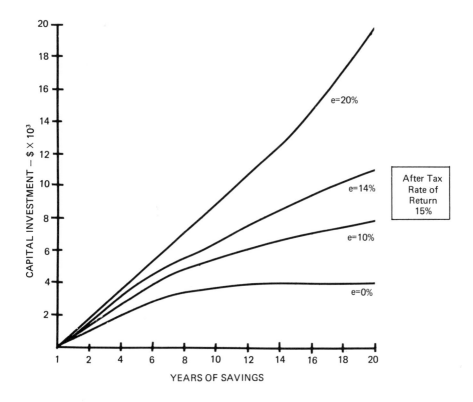

Figure 2-12. Effects of Fuel Escalation

Analysis

From Figure 2-12. For each $1000 energy savings, an invest-
ment of $3750 is justified, or $15,000 for a $4000 annual savings
when no fuel increase is accounted for.

With a 14% fuel escalation rate an investment of $8,985 is jus-
tified for each $1,000 energy savings; thus $35,940 can be justified
for $4,000 savings.

An expenditure of more than double is economically justifiable
and will yield the same after-tax rate of return of 15% when a fuel
escalation of 14% is considered.

COMPUTER PROGRAM SYSTEMS
FOR ECONOMIC ANALYSIS

The methods outlined in this chapter are easily adapted to computer algorithms. One such program which has gained wide attention utilizes a Radio Shack TRS–80 Computer in conjunction with Carrier Corporation software on program discs. For approximately $1000, software to enable life-cycle cost analysis and evaluation utilizing eight other programs is available.

The advantage of the computer system is that it is simple to use, allows input of fuel inflation rates, and comparisons of various alternates, and is relatively low in cost.

REFERENCE

1.   *Waste Heat Management Guidebook*, NBS Handbook 121, U.S. Government Printing Office, Washington, DC 20402.

3

HVAC Energy Recovery

Robert J. Kinnier*

Consideration should be given to HVAC energy recovery for both new and retrofit applications where building codes or other regulations require substantial amounts of fresh make-up air. Air-to-air heat exchangers can be used to pre-heat or pre-cool supply air from wasted heat or cooling being exhausted from the building.

In addition to the obvious fuel savings, several other benefits are possible. For instance, when designed into new installations, major savings can be achieved through the reduction in the required capacities of heating and cooling equipment. Quite often, these initial capital cost savings alone can justify energy recovery equipment. In existing manufacturing plants, relieving a negative building pressure can be as great a fuel savings as recovering waste heat from building ventilation exhaust.

For HVAC applications, four basic types of air-to-air heat exchangers are used: heat wheels, plate exchangers, heat pipes and coil loops. They are available by themselves or incorporated into factory built packaged energy recovery systems with supply and exhaust fans, filters, supplementary heating and cooling coils, electrical and control wiring. Costs for a complete basic heat wheel system can range from $2.50/cfm down to $.70/cfm, depending on the size of

*Robert J. Kinnier's material used in this chapter first appeared in *Energy Engineering*, Vol. 78, No. 2, entitled "HVAC Energy Recovery Bin Method and Life-Cycle Costing."

the unit. See Figure 3-1. Most of the applications in this chapter fall within the low to medium temperature heat recovery range.

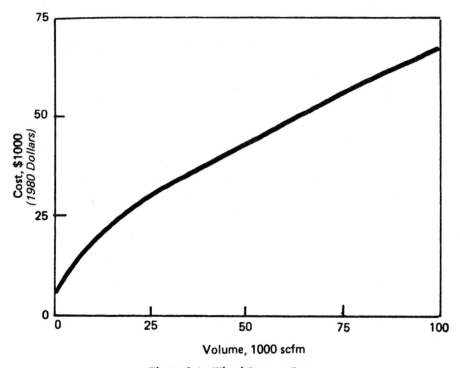

Figure 3-1. Wheel System Cost

Energy recovery equipment costs should be evaluated on an incremental basis. This means isolating those costs incurred by using energy recovery from those costs which would have been incurred anyway. For example, if energy recovery was not used in a new building, a make-up air handling unit including a fan, coil, filter, electrical and controls would still be required, as well as some type of exhaust fan. The evaluation form shown in Figure 3-2 can be used to calculate the incremental cost of energy recovery.

Since annual HVAC fuel savings utilizing an air-to-air exchanger depends on the temperature difference between outside and exhaust air from the building, climatic conditions of the building location must be considered. This requires that frequency and range of outside

Energy recovery system cost	$ [ ]
Plus extra ductwork	$ [ ]
Less reduction in heating and cooling equipment sizes	$ ( )
Less basic HVAC system components	$ ( )
Incremental energy recovery cost	$ [ ]

Figure 3-2. Incremental Cost Evaluation

air temperature variation be calculated for the period that the HVAC system is operating. Two methods are available: the *ASHRAE modified degree day procedure* and *engineering weather data bin method.*

The degree day procedure is based on averaging and therefore does not take into account the time of day when the system operates. Also, it provides cooling hours to calculate cooling energy savings. In addition, the degree day procedure does not provide wet bulb temperature data to calculate latent heat transfer.

The engineering weather data bin method requires cumbersome calculations but provides a more accurate analysis of energy savings. As shown in Figure 3-3, the bin method provides the hourly frequency of occurrence in 5F dry-bulb temperature increments and mean coincident wet-bulb temperatures (MCWB) for each day divided into three time periods.

The first step in calculating energy savings is to determine the efficiency of the air-to-air exchanger from the manufacturer's published data. In most HVAC system applications, particularly where air conditioning is required, it is advantageous to recover both sensible and latent heat. Building air conditioning systems requiring fresh air make-up during the summer actually have more latent heat available for recovery than sensible heat, as shown in Figure 3-4.

Temperature Range	FEBRUARY				MARCH					APRIL					ANNUAL TOTAL				
	Obsn Hour Gp 09 to 16	17 to 24	Total Obsn	M C W B	01 to 08	09 to 16	17 to 24	Total Obsn	M C W B	01 to 08	09 to 16	17 to 24	Total Obsn	M C W B	01 to 08	09 to 16	17 to 24	Total Obsn	M C W B
100/104																		0	77
95/99																6	0	6	76
90/94																52	6	58	74
85/89											1	0	1	65	1	132	32	165	72
80/84						0		0	62		4	1	5	64	11	225	88	324	70
75/79						0		0	61		9	3	12	63	68	260	159	487	67
70/74						2	0	2	57	1	18	8	27	60	176	262	243	681	64
65/69	0		0	60	0	3	1	4	55	6	19	11	36	57	270	220	269	759	61
60/64	1	0	1	55	0	5	3	8	53	13	24	18	55	54	280	176	244	700	57
55/59	2	1	4	52	3	9	5	17	50	14	25	22	61	50	233	160	211	604	52
50/54	4	2	7	47	4	15	8	27	46	25	33	32	90	46	222	163	196	581	47
45/49	8	4	14	43	10	22	14	46	42	35	39	36	110	42	212	164	189	565	43
40/44	18	10	32	38	15	39	27	81	37	44	37	49	130	38	194	181	197	572	38
35/39	42	30	90	34	41	57	51	149	34	50	23	35	108	34	243	237	245	725	34
30/34	57	61	171	30	72	50	69	191	30	33	7	20	60	30	319	251	299	869	30
25/29	37	46	128	25	41	24	36	101	25	14	1	5	20	25	227	160	202	589	25
20/24	21	28	79	21	31	14	17	62	21	4	0	1	5	20	154	100	117	371	21
15/19	13	16	52	16	16	5	9	30	16	1	0		1	16	98	60	73	231	16
10/14	9	8	33	11	9	1	4	14	11						69	43	52	164	11
5/9	5	8	22	6	3	0	1	4	6						44	30	41	115	6
0/4	5	6	19	1	3	0	0	3	2						40	18	31	89	1
-5/-1	2	4	13	-4	0	0		0	-3						27	9	17	53	-3
-10/-6	0	1	6	-8	0			0	-7						17	3	7	27	-8
-15/-11	0	0	1	-13											8	1	2	11	-13
-20/-16	0			-16											2	0	0	2	-17
-25/-21																		0	-21

Figure 3-3. Excerpt from Weather Data for Chicago (O'Hare Airport) from *Engineering Weather Data*, July 1, 1978, published by Dept. of Defense

Since heat wheels are the only air-to-air exchanger which recovers both sensible and latent heat, they are generally the most cost effective energy recovery device for HVAC systems. In addition to preventing moisture from entering the building during the summer, heat wheels also prevent moisture from being exhausted in winter. As a result, the need for humidification is reduced and the level of comfort increased. Shown in Figure 3-5 are typical performance data for a heat wheel.

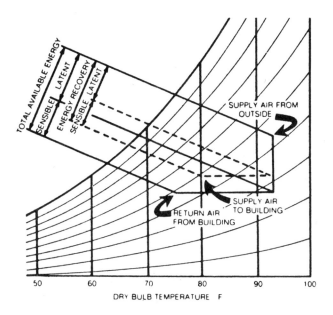

Figure 3-4. Total Energy Recovery

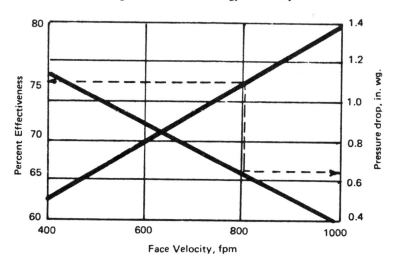

Figure 3-5. Wheel Efficiency (Sensible and Latent)
and Pressure Drop

CALCULATION OF ENERGY SAVINGS

The energy savings analysis should next be segmented into three modes of HVAC system operation: heating, ventilating and cooling. In the heating mode, energy savings are calculated on the basis of either sensible or latent heat, depending on whether the building is being humidified. If the building is humidified in the winter, the energy savings should be on the basis of latent heat. If it is not humidified, they should be on the basis of sensible heat. In the ventilating mode, there are no energy savings, since fresh air is brought in without tempering. The fan operating expense to handle the exchanger pressure drop during the ventilating mode should be determined. If a heat wheel is used, the cooling mode energy savings should be based on latent heat recovery.

Winter Energy Savings, No Humidification

After efficiency is determined, the next step is to calculate the dry-bulb temperature of the air leaving the exchanger and supplied to the building:

$$t_2 = t_1 + E(t_3 - t_1) \qquad \qquad 3\text{-}1$$

where:

t_2 = supply air temperature to building, F_{db}
t_1 = outside temperature, F_{db}
t_3 = exhaust temperature from building, F_{db}
E = exchanger effectiveness, percent

This calculation is made for the average dry-bulb temperature for each bin, i.e., 25/29F is 27F. Sensible heat recovered and returned to the building is then calculated for each bin:

$$Q_s = 1.08 \, (V_1) \, (t_2 - t_1) \, (H) \qquad \qquad 3\text{-}2$$

where:

Q_s = sensible heat recovered annually, Btu
V_1 = supply air volume, scfm
H = annual hours of occurrence

The bins are summed up to determine the total winter Btu savings. The maximum outside temperature limit for winter recovery is determined when outside air temperature equals desired supply air temperature to the building or the exhaust air temperature from the building. For heating mode only applications, it is most cost effective to stop the energy recovery system when the fuel savings drop below fan operating expense.

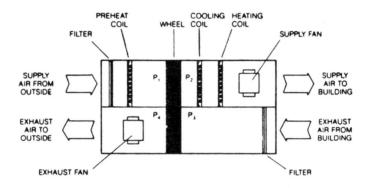

System Flow Schematic

Winter Energy Savings, With Humidification

For buildings which are humidified in winter, calculate supply air enthalpy leaving the exchanger to the building for each bin:

$$h_2 = h_1 + E(h_3 - h_1) \qquad\qquad 3\text{-}3$$

where:

h_2 = supply air enthalpy to building, Btu/lb
h_1 = outside air enthalpy, Btu/lb
h_3 = exhaust air enthalpy, Btu/lb

Total sensible and latent heat recovered and returned to the building is then:

$$Q_T = 4.5(V_1)(h_2 - h_1)(H) \qquad\qquad 3\text{-}4$$

where:

Q_T = total sensible and latent heat recovered annually, Btu

Btu's recovered in winter are converted to annual dollars saved (S_w) based on cost of fuel and efficiency of the heating system:

$$S_w = \frac{(Q_s \text{ or } Q_T) (C_F)}{e} \qquad \qquad 3\text{-}5$$

where:

S_w = annual heating savings, dollars
C_F = cost of fuel, $/Btu
e = heating equipment efficiency, percent

Energy recovery equipment fan power expense must be considered. The additional annual fan power expense (P_E) required to overcome the pressure drop across the air-to-air exchanger is a function of total annual hours the system operates (H_T):

$$P_E = 0.000147 \; [(V_1) (\Delta P_1) + (V_3) (\Delta P_3)] \; (H_T) (C_E) \qquad 3\text{-}6$$

where:

P_E = annual fan power expense, dollars
ΔP_1 = supply pressure drop across exchanger, in. wg
V_3 = exhaust volume, scfm
ΔP_3 = exhaust pressure drop across exchanger, in. wg
H_T = annual hours system operates
C_E = cost of electricity, $/kwh

Ventilation Expense

Fan power expense can be calculated individually for each mode of operation using Formula 3-6 or on the basis of total annual hours which the system operates. If the supply and exhaust air pass through the exchanger during the ventilation mode, even though no energy is recovered, the additional fan power expense should be considered.

Summer Energy Savings

Since air conditioning usually requires dehumidification of outside air, a heat wheel capable of latent recovery is particularly beneficial for summer cooling applications. By condensing outside air moisture on the heat wheel, transferring it to the exhaust air stream by rotating the media, the required cooling capacity of the refrigeration coil is significantly reduced. To calculate energy savings when the exchanger can recover latent heat, use Formulas 3-3 and 3-4. The bins are summed up, starting where the outside air temperature exceeds the required supply air temperature to the building. Btu's recovered in summer are converted to annual dollars saved (S_S) based on the cost of electricity and the cooling power input factor:

$$S_S = 0.0000833 \ (Q_T)(C_E)(P_C) \hspace{2cm} 3\text{-}7$$

where:

S_S = annual cooling savings, dollars

P_C = cooling power input, kwh/ton of refrigeration. See Table 3-1.

Additional fan power expense can be calculated from Formula 3-6.

Table 3-1. Cooling Power Input Data for Formula 3-7

Condenser Cooling Media	Refrigeration, tons	Cooling Power, kwh/ton
Air	100	1.40
Water	25–100	1.11
Water	>100	0.99

To obtain net operating savings using energy recovery, maintenance and insurance as well as the fan power expense should be deducted from fuel savings. As a rule of thumb, annual maintenance and insurance costs can be taken at 1½% of installed equipment cost. This expense should also be considered on an incremental basis, as previously discussed.

$$Net\ operating = \frac{Fuel}{savings} = \frac{Electric}{power} - \frac{Maintenance}{\&\ Insurance} \qquad 3\text{-}8$$

SIM 3-1

A new commercial office building in Chicago is designed to exhaust 13,000 cfm at 75F from 6 a.m. to 6 p.m., 5 days a week, summer and winter. The building is not humidified during the winter. Indoor humidity is expected to average 50% rh during the summer. Current energy rates are $0.105/kwh for electricity and $0.29/therm for natural gas.

Analysis

From Figure 3-1, energy recovery equipment cost is determined to be $23,000. Evaluating the project on an incremental cost basis, assume that it will cost $5,000 extra for special ductwork to bring the supply and exhaust air streams together. To provide a conservative design, credit is taken only for the reduction in the cooling load at $240/ton. In this case, energy recovery will result in a $9,700 equipment cost savings.

If energy recovery was not used on the project, a HVAC penthouse unit at 70¢/cfm and exhaust fan at 15¢/cfm would be required. It is also assumed that installation costs would be approximately the same for either energy recovery or a penthouse unit and exhaust fan.

The resulting incremental cost (Figure 3-2) for energy recovery on this project is:

Energy recovery system cost	$23,000
Plus extra ductwork	5,000
Less reduction in heating cooling equipment sizes	(9,700)
Less basic HVAC system components	(11,000)
Incremental energy recovery cost	$ 7,300

From Figure 3-5, a nominally selected 600 fpm heat wheel has a 75% effectiveness and 0.70 in. wg pressure drop across the media. From Formula 3-1, with no winter humidification, for the 25/29F bin, supply air temperature is

$$t_2 = 27F + 75\% (75F - 27F) = 63F$$

Since the system will be operating only 12 hours a day, 5 days a week, it is necessary to factor the weather data which are based on a 7-day week. Also, since 6 a.m. − 6 p.m. overlaps all three bins, to simplify calculations, the hours of occurrence are based on 8 hours from the 09 to 16 (8 a.m. − 4 p.m.) observation hour group and 4 hours from the 17 − 24 (4 p.m. to 12 a.m.) group. From Figure 3-3 (Annual Totals) and Formula 3-2,

$$Q_S = 1.08(13,000 \text{ scfm})(63\text{F} − 27\text{F}) \, [160 \text{ hr} + (1/2)202 \text{ hr}]$$

$$\left[\frac{5 \text{ days/week}}{7 \text{ days/week}}\right]$$

$$= 94 \times 10^6 \text{ Btu}$$

For a heating system efficiency of 80%, from Formula 3-5,

$$S_w = \frac{(94 \times 10^6 \text{ Btu}) \left(\dfrac{\$0.29}{\text{therm}}\right) \left(\dfrac{\text{therm}}{10^5 \text{ Btu}}\right)}{80\%}$$

$$= \$341$$

From Formula 3-6,

$$P_E = 0.000147 \, [(13,000 \text{ scfm})(0.7 \text{ in. wg}) \\ + (13,000 \text{ scfm})(0.7 \text{ in. wg})]$$

$$[160 \text{ hr} + (1/2)202 \text{ hr}] \left[\frac{5 \text{ days/week}}{7 \text{ days/week}}\right]$$
$$(\$0.105/\text{kwh})$$

$$= \$52$$

Fuel savings for the 80/84F bin in the cooling mode are calculated based on a 1.40 (air cooled condenser) cooling power input factor; from Formula 3-3,

$$h_2 = 34.25 \text{ Btu/lb} + 75\% (28.69 \text{ Btu/lb} − 34.25 \text{ Btu/lb}) \\ = 30.08 \text{ Btu/lb}$$

From Formula 3-4,

$$Q_T = 4.5(13,000 \text{ scfm}) (30.08 \text{ Btu/lb}-34.25 \text{ Btu/lb})$$

$$[225 \text{ hr} + (1/2) 88 \text{ hr}] \left[\frac{5 \text{ days/week}}{7 \text{ days/week}}\right]$$

$$= -47 \times 10^6 \text{ Btu}, \text{ where the negative sign indicates cooling saved.}$$

From Formula 3-7,

$$S_S = 0.0000833 (47 \times 10^6 \text{ Btu}) \left(\frac{\$0.105}{\text{kwh}}\right) \left(\frac{1.40 \text{ kwh}}{\text{ton}}\right)$$

$$= \$576$$

Figure 3-6 is a summary of calculations for all bins. Since the building is not humidified in the winter, annual heating savings are based on sensible heat recovery only, or 778 million Btu. The heat wheel, however, also saves an additional 190 million Btu, or 24% extra, of latent heat in the winter, because it prevents moisture from leaving the building.

Net operating savings in this example are determined by deductting maintenance and insurance from the annual savings:

Maintenance & Insurance = $7,300 × 1½% = $110

Net operating savings = (2827 + $1609) − $877 − $110
= $3449/year

To determine if this is a good investment, a simple payback technique would be used:

$$Payback = \frac{Installed \ equipment \ cost}{Net \ operating \ savings}$$

$$= \frac{\$7,300}{\$3,449/\text{year}}$$

$$= 2.12 \text{ years}$$

Although a 2.12 year payback will be considered attractive, this analysis fails to consider inflation as well as the timing of cash flows. The significance of rising energy costs is illustrated in Figure 3-7. For example, if fuel escalates at 10% per year and the cost of capital

HVAC System Mode	Outside Temperature		Annual Hours of Occurrence		Supply Temperature	Annual Heat Recovered 10^6 Btu		Annual Savings		Annual Fan Power Expense
	°FDB	°FWB	8 AM-4 PM	one half 4 PM-12 AM	T_2, °FDB	Q_s	Q_t	Heating	Cooling	
Cooling	95/99	76	6	0	81		2		$ 26	$ 1
	90/94	74	52	3.0	79		16		193	11
	85/89	72	132	16.0	78		34		416	30
	80/84	70	225	44.0	77		47		576	54
	75/79	67	260	78.5	76		32		398	68
Ventilation	70/74	64	262	121.5	72	—	—	—	—	77
Heating	65/69	61	220	134.5	73	21		$ 77		71
	60/64	57	176	122.0	72	29		106		60
	55/59	52	160	105.5	71	36		130		53
	50/54	47	163	98.0	69	45		164		52
	45/49	43	164	94.5	68	54		197		52
	40/44	38	181	98.5	67	69		252		56
	35/39	34	237	122.5	66	103		373		72
	30/34	30	251	149.5	64	130		470		81
	25/29	25	160	101.0	63	94		341		52
	20/24	21	100	58.5	62	63		229		32
	15/19	16	60	36.5	61	42		153		19
	10/14	11	43	26.0	59	33		119		14
	5/9	6	30	20.5	58	26		94		10
	0/4	1	18	15.5	57	18		67		7
	-5/-1	-3	9	8.5	56	10		37		4
	-10/-6	-8	3	3.5	54	4		15		1
	-15/-11	-13	1	1.0	53	1		3		0
TOTAL	—	—	—	—	—	778	131	$2827	$1609	$877

Figure 3-6. Summary of Calculations for Chicago Commercial Office Building

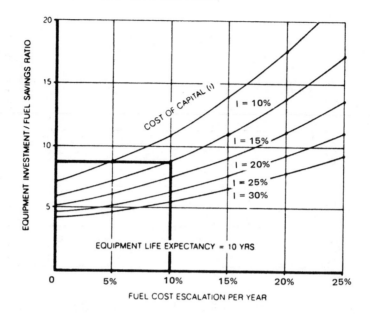

Figure 3-7. Energy Recovery Investment

is 15%, for every $1 of fuel saved today, $9 can be expended for energy recovery equipment.

To evaluate economic alternatives more accurately, life cycle costing is used. Life cycle costing combines the impact of fuel, electrical and labor inflation with taxes, depreciation and tax credits to determine the return on investment. Additionally, an incremental costing approach should be used. For instance, if an energy recovery system is being considered as a replacement for a worn out air handler, the incremental cost difference between a new air handler and energy recovery system should be used as the basis of the life cycle cost analysis.

In this example, consider the impact of 10% annual inflation for gas and 8% for electric power plus 7% annual inflation for maintenance and insurance as shown in Figure 3-8.

In addition to inflation, financial factors such as depreciation, taxes and tax credits must also be considered. These factors depend on the type of energy recovery application as shown in Figure 3-9.

Base Year Inflation Rate		*Fuel Savings*		*Electric Power $−877 8%*	*Maintenance & Insurance $ −110 7%*	*Net Operating Savings*	
		Gas $2827 10%	*Electric $1609 8%*				
YEAR	1	$3110	$1738	$ −947	$ −118	$3783	1
	2	3421	1877	−1023	−126	4149	2
	3	3763	2027	−1105	−135	4550	3
	4	4139	2189	−1193	−144	4991	4
	5	4553	2364	−1289	−154	5474	5
	6	5008	2553	−1392	−165	6005	6
	7	5509	2758	−1503	−177	6587	7
	8	6060	2978	−1623	−189	7226	8
	9	6666	3216	−1753	−202	7927	9
	10	7333	3474	−1893	−216	8696	10

Figure 3-8. Incremental Operation Savings/Expenses

Application	Type Construction	Financial Consideration			
		Taxes	Depre-ciation	Investment Tax Credit	Energy Tax Credit
Commercial	Retrofit[1]	X	X		X
Industrial HVAC	New	X	X		
Gov't, Charitable	Retrofit				
Institution HVAC	New				
Industrial	Retrofit[1]	X	X	X	X
Process	New	X	X	X	

Note 1: HVAC systems and industrial processes existing prior to October 1, 1978.

Figure 5-9. Financial Considerations

The Energy Tax Act of 1978 allows a 10% *energy* tax credit for retrofitting energy recovery equipment to commercial or industrial HVAC systems as well as industrial processes existing prior to October 1, 1978. If the equipment is used in conjunction with a new or existing industrial process, a 10% *investment* tax credit applies. Retrofit industrial processes are therefore entitled to a 20% total tax credit for application to processes existing prior to October 1, 1978.

Income taxes for commercial office buildings must be paid on the net operating savings. The federal income tax rate for corporations is approximately 48%. Income taxes are partially offset by depreciating the installed cost of the energy recovery system:

$$Taxes = Tax\ rate\ (Net\ operating\ savings - Depreciation) \qquad 3\text{-}9$$

Sum-of-the-years-digits depreciation can be used for energy recovery equipment, since it provides the earliest depreciation allowable by the Internal Revenue Service and consequently, a more attractive return on investment. Deducting taxes from the net operating savings, we obtain the net cash flow:

$$Net\ cash\ flow = Net\ operating\ savings - Taxes \qquad 3\text{-}10$$

For this example, assume that a 48% tax rate applies and that it is new construction; then cash flow will be as shown in Figure 3-10. Since the net cash flow will be occurring over a ten year period, it is

Year	Net Operating Savings	Sum-of-the-Years-Digits Depreciation	Taxes @ 48%	Energy Tax Credit	Net Cash Flow	
Acquisition	$ -6981	$ 664	$ 319	—	$ -6981	
1	2572	1261	-1210		2572	1
2	2699	1128	-1450		2699	2
3	2844	995	-1706		2844	3
4	3009	863	-1981		3009	4
5	3197	730	-2277		3197	5
6	3409	597	-2596		3409	6
7	3648	465	-2939		3648	7
8	3917	332	-3309		3917	8
9	4218	199	-3709		4218	9
10	4554	66	-4142		4554	10

Figure 3-10. Tax and Depreciation Impact

necessary to consider the time value of money using present value analysis. Simply stated, $1 saved 10 years from now is worth considerably less than $1 in the pocket today. The reason being, that the $1 in hand today could be put in the bank at a compounded interest rate and be worth considerably more in ten years. The today's $1 depends on the period of time and the interest rate. Conversely, $1 saved in the future, can be discounted to its present value in the same manner:

$$Present\ value = \frac{Future\ value}{(1 + i)^n} \qquad\qquad 3\text{-}11$$

where i = annual interest rate, percent
n = period, year

In this life cycle costing analysis, the annual interest rate (i) is really the after-tax return on investment (ROI). Determining ROI is a trial-and-error procedure which equates installed equipment cost to the sum of the present value of net cash flow:

$$Installed\ equipment = \sum_{n=1}^{10} \frac{F}{(1 + i)^n} \qquad\qquad 3\text{-}12$$

where F = future net operating savings, in dollars, for period (n).

For this example, (i) or (ROI) is assumed to be 40%. Present value net cash flow is given in Figure 3-11. In this case, ROI is slightly over 40%. Whether the ROI is an attractive investment is determined by the individual company's expected return on its capital. Typical returns on capital for various industries are shown in Figure 3-12.

One other factor must be considered to determine whether or not the ROI is attractive: risk. Actual after-tax ROI's shown in Figure 3-12 include risk or, in other words, the end result of projects which succeeded and others which failed. The typical minimum after-tax ROI for HVAC energy recovery projects is 15%.

A 40% after-tax return would be considered a very attractive return on investment for any industry. In addition, *qualitative* bene-

Year	Present Value Net Cash Flow	$\frac{1}{(1+i)^n}$ $i = 40\%$	Future Value Net Cash Flow By Year (n)									
			1	2	3	4	5	6	7	8	9	10
Acquisition	$-6981	1.0000										
1	1837	0.7143	$2572									
2	1377	0.5102		$2699								
3	1036	0.3644			$2844							
4	783	0.2603				$3009						
5	594	0.1859					$3197					
6	453	0.1328						$3409				
7	346	0.0949							$3648			
8	265	0.0678								$3917		
9	204	0.0484									$4218	
10	157	0.0346										$4554
Sum of Present Value	$ 71											

Figure 3-11. Present Value Net Cash Flow

Industry	Industry Median
Personal Products	13.6%
Aerospace & Defense	11.2%
Information Processing	11.2%
Leisure	12.4%
Health Care	13.6%
Electronics & Electrical Equipment	11.6%
Energy	11.3%
Wholesalers	12.5%
Chemicals	11.3%
Construction — Contractors	11.0%
Industrial Equipment	12.0%
Forest Products & Packaging	10.4%
Utilities — Natural Gas	7.7%
Food & Drink	10.5%
Finance — Insurance	13.2%
Automotive	10.0%
Finance — Consumer Finance	7.9%
Supermarkets	8.8%
Multicompanies — Conglomerates	8.8%
Multicompanies — Multi-Industry	9.7%
Household Products	9.2%
Retail Distribution	9.4%
Finance — Banks	10.4%
Construction — Building Materials	8.7%
Utilities — Electric & Telephone	6.2%
Transportation — Surface	6.5%
Metals	7.5%
Steel	8.4%
Apparel	8.7%
Transportation — Air	5.9%
All Industry Medians	10.2%

Source: *Forbes Magazine,* January 8, 1979

Figure 3-12. Typical Returns After Tax on Total Capital

fits can be equally important. For instance, relieving high negative building pressure can reduce perimeter heating costs and improve productivity by eliminating employee discomfort due to drafts.

Fuel curtailments are another important consideration. If shortages, curtailments or quotas for fuels occur, reduced consumption obtained through the use of an HVAC air-to-air exchanger can be the difference between operating, curtailing, or shutting down a plant. In some cases, HVAC energy recovery may be the only means of overcoming the fuel allocation limit in a building expansion.

4

Waste Heat Recovery from Combustion Devices

Boilers, incinerators and process furnaces represent potential sources for recovering waste heat.

BACKGROUND

Waste heat recovery from combustion devices are generally heat exchangers that recover the waste heat content from the exhaust flue gas to

1. preheat combustion air—air preheater
2. heat boiler feedwater—economizer
3. produce steam—waste heat boiler
4. produce electricity—heat engines
5. produce hot water—condensation heat recovery

Air preheaters, economizers and waste heat boilers have been commercialized for a number of years and are considered standard technology. However, existing technology has considered a minimum exhaust temperature of 270-300F due to the formation of acidic condensate.

Heat engines are discussed in Chapter 5 and condensation heat recovery is presented in Chapter 7. This chapter will discuss two types of waste heat recovery devices that allow for exhaust gas temperatures of less than 250F; Deep Economizers—exhaust gas temperatures of 150–160F, and Indirect Contact Condensation Heat Recovery—exhaust gas temperatures of 100–110F.

EFFICIENCY INCREASE

The efficiency increase potential from a combustion device is determined by the exhaust gas temperature and excess air rate. Figure 4-1 shows a general relationship to estimate the efficiency increase from each 10F drop in exhaust gas temperature. This relationship is only valid for exhaust gas temperature greater than 150F.

SIM 4-1

A boiler has a flue gas temperature of 475F and excess air rate of 30%. Estimate the efficiency increase from installing a heat recovery device that reduces the exhaust gas to 300F.

Efficiency increase per 10F temperature drop = 0.27 (Fig. 4-1)
Total temperature drop = 475–300 = 175F
Efficiency increase = 175F/10F X .27%/F = 4.7%

SIM 4-2

Estimate the efficiency increase for the same boiler with a heat recovery device exhaust gas temperature of 150F.

Efficiency increase per 10F temperature drop = 0.27%/F
Total temperature drop = 475–150 = 325F
Efficiency increase = 325 F/10F X 0.27%/F = 8.8%

The increase in efficiency improvement from 4.7% to 8.8% is the justification for designing a device that is suitable for low exhaust gas temperatures.

DEEP ECONOMIZERS

Deep Economizers are economizers that are designed to handle the acidic condensate that results from cooling a flue gas below 270F. The primary design variable is the material of construction of the tubes at the cold end of the device. Typical systems design are:

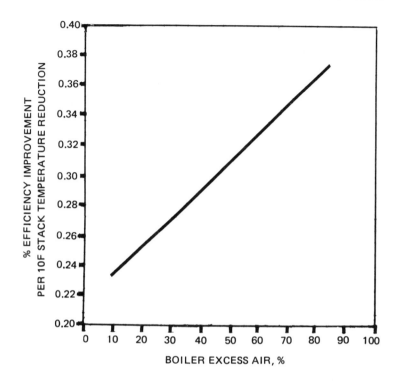

Figure 4-1. Percent Efficiency Improvement for Every 10F Drop in Stack Temperature

Valid for estimating efficiency improvements on typical natural gas, No. 2 through No. 6 oils and coal fuels.

1. Carbon steel tubes with a throwaway section at the cold end. These systems are designed with modular sections at the cold end that are easily removed and replaced on a period basis.

2. Stainless steel tubes that withstand the corrosive environment. These are standard economizers with stainless steel tubes.

3. Carbon steel tubes for the bulk of the exchanger and stainless steel tubes for the cold end. These systems have carbon steel tubes for the main section of the economizer with stainless steel only for the cold end section.

4. Glass tubed heat exchangers. These systems use glass tubes. They have been applied most extensively in gas–gas service as air preheaters. Applications with gas–liquid systems are under development.

5. Teflon tubes. These systems use Teflon tubes to withstand the corrosive environment. CHX and duPont have developed the unit and have solved the critical sealing problems that usually result in applying Teflon tubes in heat exchangers. Exhaust gas temperatures geater than 500F require two-stage systems.

The choice of compatible materials of construction for a particular project is determined primarily by the flue gas composition. However, each of the above options can be adapted to most cases. While Teflon or glass would be most suitable for applications with very corrosive gases containing high concentrations of sulfuric and hydroflouric acids, Inconel or other high alloys can be substituted for the stainless steel sections or a shorter replacement period considered for the throwaway type units.

A critical feature of the use of deep economizers is a suitable heat sink to cool the exhaust gas to the 150–160F range. Typical heating plant condensate return systems operate at 150 to 160F and usually the combined flow of cold make-up and condensate return provides for a suitable cold inlet temperature of 130 to 140F. Process furnaces and incinerators do not usually have cold heat sinks available for in-process use and other heating requirements are considered. Typical applications include, but are not limited to

1. Space heat
2. Domestic or process hot water
3. Heat engine

SIM 4-3

A hospital facility has a 15 MMBtu/hr natural gas-fired boiler operating with an average load of 10 MMBtu/hr, stack temperature of 450F and excess O_2 of 3% for the winter season. The summer season load is 5 MMBtu/hr average. Estimate the efficiency improvement, yearly fuel savings and estimated payback for a "deep economizer" operating with a 150F exit gas temperature.

Table 4-1 shows the ASME efficiency calculation for this case indicating an operating efficiency of 75.5%. The heat recovered can be calculated from the formula

$$Q, \text{Btu/hr} = w \times C_p \times (t_i - t_o) \qquad \text{4-1}$$

where Q = heat recovered, Btu/hr
w = lbs per hour of dry gas
C_p = specific heat of dry gas (modified for water content)
     = .24 + ($h \times$ .45) Btu/lb/F
t_i = inlet temperature, F
t_o = outlet temperature, F
h = humidity, lbs of water/lb of dry gas

The dry gas flow rate is calculated from the formula:

$$w = N \times (Q_i/E) / HHV \qquad \text{4-2}$$

where w = lbs per hour of dry gas
N = lbs of dry gas/lb of fuel
Q_i = energy output from boiler, MMBtu/hr
E = operating efficiency, %
HHV = higher heating value of fuel, Btu/lb

For this example

w = 20.52 lbs dry gas/lbs fuel $\times$ (10 $\times$ 10^6 Btu/hr/.755)
     $\times$ 22500 Btu/lb
w = 12080 lb/hr

The humidity is estimated as

h = 9 $\times$ %H/N
h = .11 lb water/lb dry gas

The specific heat, C_p, is

$$C_p = .24 + .11 \times .45 = .29 \text{ Btu/lb/F}$$

The heat recovery rate is calculated from Formula 4-1

$$Q = 12080 \times .29 \times (450 - 150) = 1.05 \text{ MMBtu/hr}$$

Table 4-1. ASME Heat Loss

	Boiler Fuel	Gas	
	Nitrogen, %	88.00	
	Carbon/lb	0.75	
	Dry Gas/lb	20.53	
	Excess Air, %	14.57	

Losses	Btu/lb	%
Dry Gas	1823	8.10
Latent	2740	12.18
Moisture	0	0.00
Refuse	0	0.00
Radiation	609	2.71
Miscellaneous	338	1.50
TOTAL	5509	24.48

	Efficiency, %	75.52
	Fuel Flow Rate, lb/H	589

The efficiency increase is

$$\text{Efficiency increase} = [1.05 \, / \, (10/.755)] \times 100 = 7.9\%$$

In this application the hot water load is constant and comprises greater than 7.9% of the load during all circumstances. Therefore, the required heat sink in the form of cold inlet water is available and the recovered heat can be utilized all year. Therefore, the estimated savings is based on the yearly fuel consumption.

Savings/year = Yearly fuel consumption $\times$ (% recovery/efficiency) $\times$ cost per MMBtu $-$ operating & maintenance costs

Savings/year = 65,520 MMBtu/yr $\times$ 7.9/75.5
$\times$ \$5.00/MMBtu$-$\$8000/yr
= \$26,280/year

The cost for an economizer project can be estimated at $50,000 per million Btu's recovered. The cost for this example is approximately $50,000 and the simple payback period is 1.9 years.

For a more detailed look, Table 4-2 shows the monthly cash flow based on a 2 year, 25% interest capitalization with a 47% income tax rate. The after-tax savings/year is $14,300/yr and the return on investment for a 5 year period is 17% after taxes.

INDIRECT CONTACT CONDENSATION HEAT RECOVERY

Indirect Contact Condensation Heat Recovery is the term used for standard type shell and tube heat exchangers that are designed to cool the flue gas into the condensing region. The effect on heat recovery rates is quite dramatic as indicated in Figure 4-2. Cooling the flue gas from 300F to 140F results in a 3% efficiency increase. Further cooling results in condensation of water vapor and the heat recovery rate increases rapidly. At 100F exit temperature, the efficiency increase is 12%.

The same materials of construction requirements as described for the "deep economizers" are applicable. Stainless steel, glass and Teflon have all been used. Costs range from $50,000/million Btus recovered for stainless steel to $75,000/million Btus for Teflon units.

Energy Recovery

The energy recovery rates in a condensing heat exchanger can be calculated as follows:

$$Q = w \left[I(i) - I(o) \right] \qquad 4\text{-}3$$

where Q = heat recovered, Btu/hr
w = dry gas flow rate, Lbs/hr
I = enthalpy of dry gas (including moisture), Btu/lb of dry gas

The heat content of the gas stream is calculated by:

Inlet $\qquad I(i) = 0.24 \times t_i + H(i) \times (.45 \times t_i + 950) \qquad 4\text{-}4$

Table 4-2. Economic Analysis

				Loan Period, Months		24
	Operating Expense & Cash Flow Summary			Interest Rate, %/Yr		25
	Monthly, $			Income Tax Rate, %		47

Month	Operating Expense	Maintenance Supervision, Insurance, Taxes & Interest	Net Savings	Net Savings (After Taxes)	Loan Payment	Cash Flow
					18.737	
1	312	966	3,366	2,176	2,668.6	−493
2	312	966	3,366	2,176	2,668.6	−493
3	312	966	3,366	2,176	2,668.6	−493
4	106	966	500	657	2,668.6	−2,012
5	106	966	500	657	2,668.6	−2,012
6	106	966	500	657	2,668.6	−2,012
7	106	966	500	657	2,668.6	−2,012
8	106	966	500	657	2,668.6	−2,012
9	106	966	500	657	2,668.6	−2,012
10	106	966	500	657	2,668.6	−2,012
11	312	966	3,366	2,176	2,668.6	−493
12	312	966	3,366	2,176	2,668.6	−493
Year # 1	2,298	11,590	20,332	15,476	32,023.0	−16,547
2	2,298	11,590	20,332	15,476	32,023.0	−16,547
		Income Tax Credit (10%)				5,000
				TOTAL		−11,547
3	2,298	5,660	26,262	14,311	0	14,311
4	2,298	5,660	26,262	14,311	0	14,311
5	2,298	5,660	26,262	14,311	0	14,311
		Simple Payback, Yrs		1.90		
		Return on Investment, %		17.00		

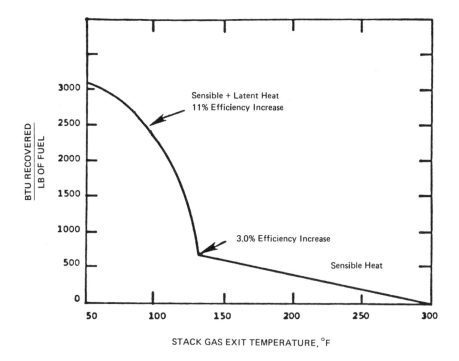

Figure 4-2. Heat Recovery Curve (Natural Gas Fired Boiler)

Outlet $\qquad I(o) = .24 \times t_0 + H(o) \times (.45 \times t_0 + 950) \qquad$ 4-5

The outlet gas is generally assumed to be saturated and the outlet humidity is calculated from equilibrium conditions. However, some manufacturers claim that surface effects in the condenser result in cold spots leading to less than saturated gas and greater latent heat recovery.

The effect of entering the condensing region can be compared with the "deep economizer" in the following simulation.

SIM 4-4

Estimate the heat recovery rate for an Indirect Condensation Heat Recovery unit with the same conditions as SIM 4-3 with an outlet gas temperature of 110F.

First calculate the heat balance:

$$I(i) = .24 \times 450 + .12 (.45 \times 450 + 950)$$
$$= 246.3 \text{ Btu/lb of dry gas}$$

$$I(o) = .24 \times 110 + .087 (.45 \times 110 + 950)$$
$$= 113.4 \text{ Btu/lb of dry gas}$$

Then

$$Q = 12080 \text{ lbs of dry gas/hr} \times (246.3{-}113.4) \text{ Btu/lb of dry gas}$$
$$= 1.61 \times 10^6 \text{ Btu/hr}$$

The energy input to the boiler = $10 \times 10^6 / .755$
$$= 13.25 \text{ MMBtu/hr}$$

Therefore, the efficiency increase = $1.61/13.25 \times 100 = 12.2\%$

The annual savings = 65,520 MMBtu/yr $\times$ 12.2/75.5
$\times$ \$5.00/MMBtu $-$ \$10,000/yr operating
& maintenance
= \$42,900/year

Using an estimated cost of \$75,000/MMBtu recovered gives a projected capital investment of \$120,800 and a simple payback period of 2.8 years.

While the condensing unit has a longer payout period than the "deep economizer," this is primarily due to the 50% increase in estimated capital cost (\$75,000 versus \$50,000/MMBtu/hr recovered). This higher cost is based primarily on the Teflon unit and a stainless steel unit could have a similar cost estimate of \$50,000/MMBtu/hr recovered which would result in an estimated capital investment of \$80,000 and a simple payback of 1.9 years. The primary criteria for use of the more expensive Teflon systems would be the tendency of the flue gas to be corrosive.

HEAT RECOVERY FOR
POLLUTION CONTROL
EQUIPMENT

Pollution control equipment typically operates at high flue gas temperatures; thus it offers a good opportunity for waste heat recovery.

One such method of meeting EPA emission requirements and recovering waste heat is by use of regenerative incineration.[1]

Solvent emission can be controlled by using a highly energy-efficient form of regenerative incineration that recovers and reuses, within the equipment, 85%, 90% or 95% of the required energy input.

Energy recovered from the solvent is used for its own incineration and fume self-destruction with very little additional fuel usage. This energy efficiency is true for spray booth cleanup as well as curing ovens. Operation isn't affected by a mixture of water-based and solvent-based exhaust streams. Equipment is simple and reliable.

Incinerators have traditionally had a bad name for fuel consumption. A modern thermal oxidation system with efficient energy recovery and recycling is different! Regenerative incineration can be tremendously efficient both in terms of cleaning the air and in terms of energy self-sustaining operation.

The regenerative system is an energy recovery unit that operates on the mass/surface area principle of heat recovery. Figure 4-3 illustrates a typical regenerative incinerator.

As hydrocarbon-laden exhaust air travels toward the central incineration chamber, it passes through a smaller chamber containing *previously warmed* energy recovery elements. The fumes, thus preheated, can be incinerated in a fraction of a second, with little or no auxiliary fuel needed in addition to the fuel value of the solvent fumes themselves.

The superheated clean air is routed through another chamber also containing the energy recovery elements. Most heat energy is stored in these energy recovery elements for use in preheating the next cycle. The minimal amount of heat carryover can be used in the plant or process. Subsequently the flow is reversed and fumes enter through the other chamber. Thus, a back and forth cycle is established

RE-THERM® OPERATION

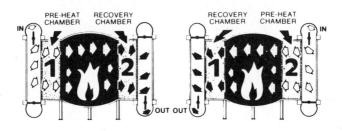

Source: REECO

Figure 4-3. Regenerative Incinerator

to capture the energy of incineration and reuse it for burning the next cycle. This feature reduces the auxiliary fuel requirement to only 5-10% of the fuel required by a common afterburner, depending on the design and allowing for wall losses. Stated another way, the regenerative incinerator can self-sustain on L.E.L. concentrations as low as 3 to 5%. Typical bake ovens are presently operated in the range of 2 to 25% L.E.L., thus a minimal amount of auxiliary fuel is required with a regenerative incinerator.

Regenerative incinerators are installed at companies such as A. E. Staley, Armstrong World Industries, Champion Spark Plug, Mead, Rexham, Western Electric, Wolverine Aluminum and others. They are controlling fumes containing a variety of otherwise troublesome items such as aluminum and steel coating fumes, plastisols, condensates, clay particulate, coffee roaster chaff, etc. Currently, there are some fifty regenerative systems in or about to be in operation. Their capacity ranges from 2,000 scfm to 60,000 scfm each. The solvent/organic concentrations range from a trace to a variable loading of from 8% to 25% L.E.L.

This energy-efficient system has a number of benefits. It offers a total system fume and odor control, and is fuel-efficient through the entire solvent load operating range. It can be installed with a minimal amount of re-work to retrofit existing lines. It eliminates complicated controllers and instrumentation as well as high maintenance tube-type preheat exchangers. Another benefit is that autoignition problems are eliminated.

Another less apparent advantage is the internal temperature profile which, to date, has prevented any condensate or particulate build-up. Nevertheless, a self-cleaning feature (which operates much like a self-cleaning oven) is incorporated.

There are many reports and evaluations regarding equipment costs, full engineering and financial studies. Such evaluations are freely available from industry sources.

A simple comparison of a current auto industry oven is also useful. The following compares an oven as presently operated, versus the same oven with a regenerative incineration system and oven feedback. The oven temperature, solvent concentrations and exhaust volume are as established by a major auto manufacturer.

To simplify this comparison, the following assumptions are used:

1. Both systems will use existing solvent-based materials.
2. There will be no basic change to the process.
3. Energy is the main cost factor and the key of the evaluation.
4. The energy cost for heating the product is the same for both systems—this energy cost may be omitted *for this comparison.*
5. The energy cost for oven wall losses, solvent vaporization, etc. is essentially the same for both—this also may be omitted for both systems.
6. When the solvents from the oven exhaust are destroyed, 70% of the resulting clean warm air can be used for oven make-up (or a boiler or plant heat or dry off, etc.).
7. The example oven exhausts 53,200 scfm @ 257F.

Using the above, a simple cost evaluation for heating the make-up air for the oven and exhausting it to atmosphere gives these results:

Oven Air Heating Cost (Figure 4-4)	$373,740/yr
Air Heating Cost with Regenerative Incineration @ 90% Thermal Energy Recovery and Feedback (Figure 4-5)	$ 84,240/yr
Savings in Air Heating Cost (Figure 4-6)	$289,500

AIR HEATING COST

Figure 4-4

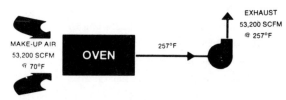

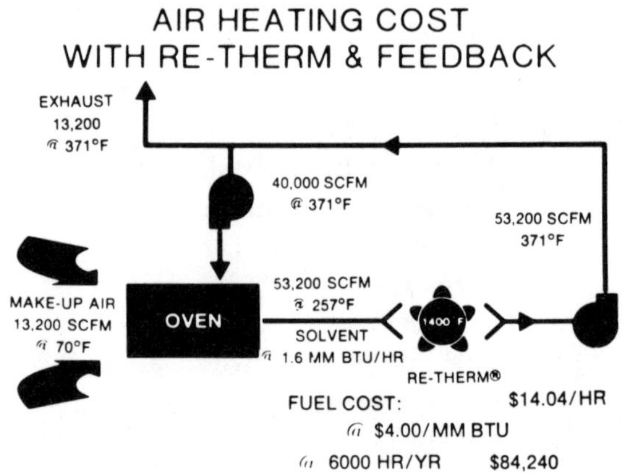

EXHAUST
53,200 SCFM
@ 257°F

MAKE-UP AIR
53,200 SCFM
@ 70°F

OVEN

257°F

FUEL COST: $62.29/HOUR
        @ $4.00/MM BTU
    @ 6000 HR/YR $373,740

AIR HEATING COST
WITH RE-THERM & FEEDBACK

EXHAUST
13,200
@ 371°F

40,000 SCFM
@ 371°F

53,200 SCFM
371°F

Figure 4-5

MAKE-UP AIR
13,200 SCFM
@ 70°F

OVEN

53,200 SCFM
@ 257°F
SOLVENT
@ 1.6 MM BTU/HR

1400 F

RE-THERM®

FUEL COST: $14.04/HR
        @ $4.00/MM BTU
    @ 6000 HR/YR $84,240

AIR HEATING COST COMPARISON

Figure 4-6

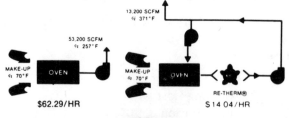

13,200 SCFM
@ 371°F

53,200 SCFM
@ 257°F

MAKE-UP
@ 70°F

OVEN

MAKE-UP
@ 70°F

OVEN

RE-THERM®

$62.29/HR $14.04/HR

SAVINGS
    @ $4.00/MM BTU $48.25/HR
    @ 6000 HR/YR $289,500/YR

The above comparison may seem simplistic. In many respects it is. The results, nevertheless, show that with existing technologies, equipment, paints and most important, existing quality, a substantial savings—energy savings—is possible with regenerative incineration.

As a case history, the Wolverine installation is a combination of seven such ovens covering four separate lines. In addition, the system provides energy to a boiler. The installation, while appearing complex, has the same simple, straightforward approach.

This system has the added advantages of allowing:

1.   Variable exhaust volume (ovens on or off)
2.   Variable solvent loads
3.   Variable energy feedback to the ovens
4.   Auxiliary boiler on or off.

The results at Wolverine are dramatic. Projected energy savings of approximately $1,000,000 per year with pollution control as a bonus.

In addition to the systems described, incinerators can be retrofitted with heat wheels, waste heat boilers, economizers and heat recovery equipment to utilize the waste heat of the exhaust gases. Refer to Chapter 8, Case Studies 8-3, 8-4, and 8-5 for additional heat recovery applications with pollution control equipment.

REFERENCE

1.   J. H. Mueller, "A Case History of Industrial Energy Conservation While Meeting EPA Regulations," *Advances in Energy Utilization Technology,* Proceedings of the 4th World Energy Engineering Congress. The Fairmont Press, Inc., Atlanta, GA 30340.

5

Low Temperature
Heat Recovery

Low temperature heat recovery technology is extremely important since there are many opportunities with this source of heat. The limit to utilizing low level heat recovery (LLHR) is that it may not be technical or economically feasible to do so.

Utilizing conventional heat recovery devices such as the heat wheel and heat exchanger for LLHR is common. LLHR as applied to HVAC was discussed in Chapter 3. This chapter will emphasize LLHR systems which use Heat Pumps, Refrigeration Systems, and Rankine Cycles.

HEAT PUMPS

The heat pump is one of the most effective devices for LLHR. It can extract low level heat, upgrade it to a suitable temperature and deliver it at the point of consumption in a very efficient manner. A heat pump can extract heat from air, water, earth, or any process fluids. The temperature range for the commercial type of heat pump is 95 to 125F. Some new industrial types of heat pumps can approach temperatures of 250F and above by utilizing special refrigerants. Heat pump applications are numerous. The following are some of the most common applications for LLHR:

- Heat recovery from ground or well water for comfort heating

- Heat recovery from building exhaust system (internal source heat pump) for transferring internal heat to the exterior of the building, or where it is most needed
- Heat recovery from low temperature solar heated water, or fluid for comfort or process heating
- Heat recovery from waste water such as sewage (effluent water), laundry discharge, commercial and industrial processes waste water

Air to Air Heat Pumps

Heat exists in air down to 460F below zero. Using outside air as a heat source has its limitations, since the efficiency of a heat pump drops off as the outside air level drops below 55F. This is because the heat is more dispersed at lower temperatures, or more difficult to capture. Thus, heat pumps are generally sized on cooling load capacities. Supplemental heat is added to compensate for declining capacity of the heat pump. This approach allows for a realistic first cost and an economical operating cost.

An average of two to three times as much heat can be moved for each Kw input compared to that produced by use of straight resistance heating. Commercially available heat pumps range in size from two to three tons for residences to up to 40 tons for commercial and industrial users. Figure 5-1 illustrates a simple scheme for determining the supplemental heat required when using an air–air heat pump.

Hydronic Heat Pump

The hydronic heat pump is similar to the air to air unit, except the heat exchange is between water and refrigerant instead of air to refrigerant, as illustrated in Figure 5-2. Depending on the position of the reversing valve, the air heat exchanger either cools or heats room air. In the case of cooling, heat is rejected through the water cooled condenser to the building water. In the case of heating, the reversing valve causes the water to refrigerant heat exchanger to become an evaporator. Heat is then absorbed from the water and discharged to the room air.

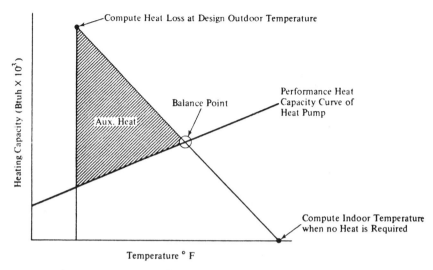

Figure 5-1. Determining Balance Point of Air to Air Heat Pump

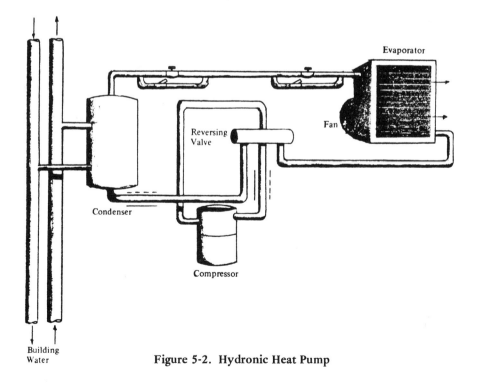

Figure 5-2. Hydronic Heat Pump

Imagine several hydronic heat pumps connected to the same building water supply. In this arrangement, it is conceivable that while one unit is providing cool air to one zone, another is providing hot air to another zone; the first heat pump is providing the heat source for the second unit, which is heating the room. This illustrates the principle of energy conservation. In practice, the heat rejected by the cooling units does not equal the heat absorbed. An additional evaporative cooler is added to the system to help balance the loads. A better heat source would be the water from wells, lakes, or rivers which is thought of as a constant heat source. Care should be taken to insure that a heat pump connected to such a heat source does not violate ecological interests.

Example of LLHR with Heat Pump[1]

The LLHR system designed for the Health and Police Science Building of the Agricultural and Technical College at Farmingdale, New York incorporates two methods to reclaim the heat from building exhaust air (see Figures 5-3 and 5-4). One method is by means of heat wheels, and the other is by means of chilled water coils as part of a heat pump transfer system. When the building is occupied, all of the air exhausted from the building is a temperature of at least 70 to 75F (much higher with heat of light). To replace this exhausted air, outside air must be brought in at temperatures which could be as low as 9F. For the perimeter areas of the building, this air must be heated in order to maintain comfort conditions at low outside temperatures. In the original design a heat recovery wheel was utilized to transfer the heat from the exhaust air to the cold outside air so that less additional heat would be required. This method of LLHR delivered an efficiency of 70% so that 9F outside air could be heated to over 40F without any external energy. Although application of the heat wheel on the exterior zones did save some energy, the poor economics of this system resulted in its elimination.

The interior areas of this building are almost always a cooling load when occupied and the direct use of cool outside air will usually help to maintain the comfort of the interior zones. The heat wheel system was not applied for the interior areas. The exhaust air from

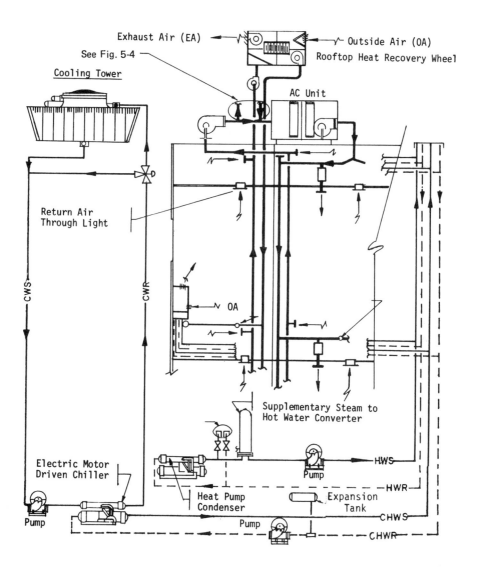

Figure 5-3. Low Level Heat Recovery System

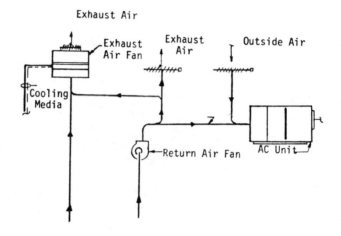

**Figure 5-4. Interior Zone Outside Air & Exhaust System
With Economizer Cycle & Exhaust Heat Pump Recovery System**

the interior carries out the heat from the building while the perimeter areas demand heat. Under these circumstances a chilled water coil can be used in the exhaust air to serve as the source of heat for a heat pump, and the heating system water can serve as the sink.

When there is no heating required in the building, all of the heat of the building is rejected through the cooling tower.

Similarly, for LLHR from waste water, a feasibility study of using effluent water as a heat source for a heat pump at one of the large water pollution control plants in New York City was completed. Analysis indicated substantial cost and energy savings due to this LLHR system.

REFRIGERATION SYSTEMS

Low grade heat from low pressure steam discharge is an ideal source of energy for single-stage absorption chillers.

Any refrigeration system uses external energy to "pump" heat from a low temperature level to a high temperature. Mechanical refrigeration compressors pump absorbed heat to a higher temperature level for heat rejection. Similarly, absorption refrigeration changes the energy level of the refrigerant (water) by using lithium bromide to alternately absorb it at a low temperature level and reject it at a high level by means of a concentration-dilution cycle.

The single-stage absorption refrigeration unit uses 10 to 12 psig steam as the driving force. Whenever users can be found for low pressure steam, energy savings will be realized. A second aspect for using absorption chillers is that they are compatible for use with solar collector systems. Several manufacturers offer absorption refrigeration equipment which uses high temperature water (160–200F) as the driving force.

A typical schematic for a single-stage absorption unit is illustrated in Figure 5-5. The basic components of the system are the evaporator, absorber, concentrator, and condenser. These components can be grouped in a single or double shell.

Evaporator. Refrigerant is sprayed over the top of the tube bundle to provide for a high rate of transfer between water in the tubes and the refrigerant on the outside of the tubes.

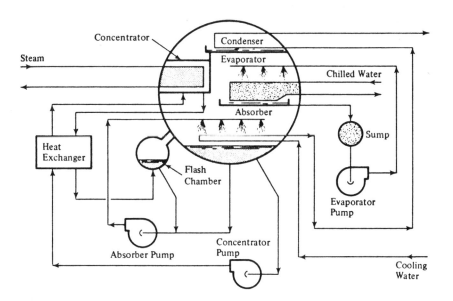

Figure 5-5. One-Shell Lithium Bromide Cycle Water Chiller
(Source: Trane Air Conditioning Manual)

Absorber. The refrigerant vapor produced in the evaporator migrates to the bottom half of the shell where it is absorbed by a lithium bromide solution. Lithium bromide is basically a salt solution which exerts a strong attractive force on the moelcules of refrigerant (water) vapor. The lithium bromide is sprayed into the absorber to speed up the condensing process. The mixture of lithium bromide and the refrigerant vapor collects in the bottom of the shell; this mixture is referred to as the dilute solution.

Concentrator. The dilute solution is then pumped through a heat exchanger where it is preheated by hot solution leaving the concentrator. The heat exchanger improves the efficiency of the cycle by reducing the amount of steam or hot water required to heat the dilute solution in the concentrator. The dilute solution enters the upper shell containing the concentrator. Steam coils supply heat to boil away the refrigerant from the solution. The absorbent left in the bottom of the concentrator has a higher percentage of absorbent than it does refrigerant, thus it is referred to as concentrated.

Condenser. The refrigerant vapor boiling from the solution in the concentrator flows upward to the condenser and is condensed. The condensed refrigerant vapor drops to the bottom of the condenser and from there flows to the evaporator through a regulating orifice. This completes the refrigerant cycle.

The single-stage absorption unit consumes approximately 18.7 pounds of steam per ton of capacity (steam consumption at full load based on typical manufacturers' data). For a single-stage absorption unit,

$$COP = \frac{1 \text{ ton} \times 12,000 \text{ Btu/ton}}{18.7 \text{ lb} \times 945 \text{ Btu/lb}} = 0.67$$

The single-stage absorption unit is not as efficient as the mechanical chiller. It is usually justified based on availability of low pressure steam, equipment considerations, or use with solar collector systems.

SIM 5-1

Compute the energy wasted when 15 psig steam is condensed prior to its return to the power plant. Comment on using the 15 psig steam directly for refrigeration.

Answer

For 30 psia steam, hfg is 945 Btu per pound of steam (from steam tables); thus, 945 Btu per pound of steam is wasted. In this case where *excess low pressure* steam cannot be used, absorption units should be considered in place of their electrical mechanical refrigeration counterparts.

SIM 5-2

4000 lb/hr of 15 psig steam is being wasted. Calculate the yearly (8000 hr/yr) energy savings if a portion of the centrifugal refrigeration system is replaced with single-stage absorption. Assume 20 Kw additional energy is required for the pumping and cooling tower cost associated with the single-stage absorption unit. Energy rate is 6¢ per Kwh and the absorption unit consumes 18.7 lb of steam per ton of capacity.

The centrifugal chiller system consumes 0.8 Kwh per ton of refrigeration.

Answer

Tons of mechanical chiller capacity replaced = 4000/18.7
$$= 213.9 \text{ tons.}$$
Yearly energy savings = 4000/18.7 × 8000 × 0.8 × $.06
$$= \$82,137$$

An example of LLHR utilizing refrigeration systems is the Hunts Point Cooperative Meat and Poultry Market in Bronx, New York. The market consists of 6.3 million cubic feet refrigerated warehouse and freezer space. As part of its energy conservation program, the market system recovers heat from the refrigeration process and utilizes it for defrosting its freezer and cooler spaces (see Figure 5-6). The plant is designed for a peak capacity of 2400 tons of refrigeration. The refrigerated calcium chloride brine is supplied at −25F and +10F in two separate circuits to freezer and cooler spaces throughout the market. The heating requirements to defrost the freezer and cooler coils is approximately 8 million Btu/hr. Ordinarily, this heating requirement would have been satisfied by the boiler

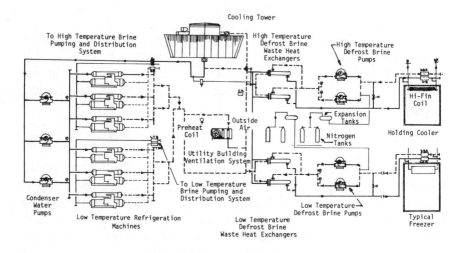

Figure 5-6. Waste Heat Reclamation Cycle

plant. It was investigated whether this heating demand could be satisfied by recapturing some of the waste heat from condenser water, normally pumped to the cooling tower for low level heat rejection. It was estimated that heat from the refrigeration process plus the heat of compression would amount to 23 million Btu/hr which is more than ample to satisfy the defrosting requirements. Based on this analysis, an LLHR system to support the defrosting system was designed (see Figure 5-6). In addition to the defrosting system LLHR also supplies heat to be used in preheating ventilation air for the utility building during winter. This LLHR was projected to save approximately 220,000 gallons/year or 20% of the peak heating requirements of the market.

RANKINE CYCLE

The Organic Rankine Cycle (ORC) converts the low level heat into usable shaft horsepower. It can utilize waste heat in the 250 to 700F range. In process industry an appreciable amount of heat is rejected around the 200 to 350F range. The conversion of this rejected heat into useful power can improve the overall thermal efficiency of a process or facility. The ORC systems as an LLHR device are presently marketed either as a packaged 500–600 kilowatts cogeneration unit or as a Binary Rankine Cycle unit to provide shaft or electric power from waste heat. Compared to the simple ORC system the Binary ORC has the advantage of being able to use high and low level heat simultaneously. Higher temperature heat (450F and above) is used in the Steam Rankine Cycle which could be open or closed, and the low level heat (200 to 250F) is used only for the closed Organic Rankine Cycle. For LLHR from diesel exhaust, industrial exhaust and steam condensate, the ORC could be an excellent device.

Hermetic Organic Rankine Cycle
(ORC) Turbine-Generator Modules

The Organic Turbine Generator is a hermetically sealed unit which contains a drive turbine, induction generator, and liquid-feed pump on a common shaft (Figure 5-7). The ORC power fluid both lubricates the bearings and cools the induction generator. Shaft seals, gears, and separate lube-oil systems are thus eliminated. This hermetic approach insures zero leakage of the power fluid.

A standard package includes the turbomachinery module, condenser, boost pump, electrical switchgear, and all required controls, plus a local control panel. The customer usually supplies the evaporators, piping, and balance of plant equipment. See Figure 5-8.

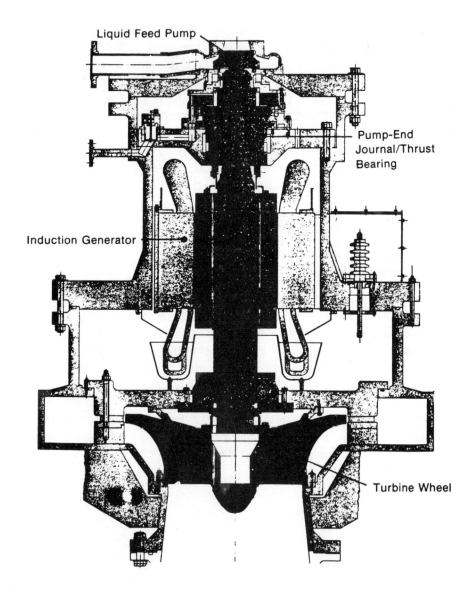

Source: Mechanical Technology Incorporated

Figure 5-7. Organic Turbine Cross Section

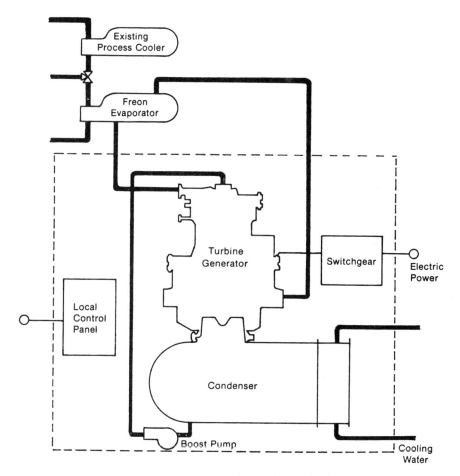

Source: Mechanical Technology Incorporated

Figure 5-8. Standard ORC Package

REFERENCE

1.   A. S. Queshi, "Low Temperature Heat Recovery," *Advances in Energy Utilization Technology*, Proceedings of the 4th World Energy Engineering Congress. The Fairmont Press, Inc. Atlanta, GA 30324.

6

High Temperature
Heat Pump Applications

The high temperature heat pump is increasingly being used to recycle waste heat in many commercial, institutional, and industrial applications. It has been coupled with solar systems to make them more efficient and has been applied as a means to a greater use of geothermal resources.

This chapter[1] concentrates on the use of these new technology heat pumps in industry, where the greatest potential for energy-saving heat recovery systems exists. Energy costs make it profitable to reclaim much of this waste heat, and with each energy price increase it is becoming more necessary for profitability.

INDUSTRIAL HEAT PUMPS

Because waste heat is found almost everywhere, high temperature commercial-industrial heat pumps can be used for a wide variety of processes on a broad spectrum of applications, or where there is no waste heat in combination with solar or geothermal sources. These heat pumps are now being used for space heating and hot water heating for industrial processes as well as for service hot water.

The major industrial uses for the heat pump appear to be in washing, blanching, sterilizing, and clean-up operations in food processing; lumber and grain drying; metal cleaning and treating processes; recycling heat in distillation operation in the food and petrochemical industries; and for industrial and commercial space and domestic water heating.

The high temperature heat pump employs the same basic Carnot cycle as a typical water chiller except that it works at higher temperatures. It offers an enormous savings potential to the large segment of business which use heat at or below 220 degrees F.

Waste heat from the source water is absorbed in the heat pump's evaporator by the unit's working fluid. The working fluid is then increased in temperature and pressure by the compressor and sent to the condenser where the heat is transferred to delivery fluid, usually water. See Figure 6-1.

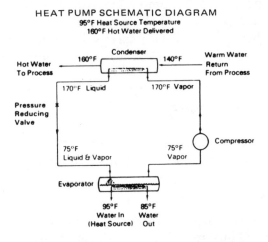

Figure 6-1. High Temperature Heat Pump

These heat pumps have coefficients of performance which typically vary between three and six, depending on the temperature amplification. This being the temperature difference between the leaving hot water from the condenser and the source water leaving the heat pump evaporator. For many typical applications it averages 4.5.

From an energy conservation standpoint, this is significant. For example, a heat pump with a COP of 4.5 uses only 0.22 units of heat input in the form of electrical energy to deliver one unit of useful heat output. The remaining 0.78 units come from the waste heat source.

Contrast this with a fossil fuel fired boiler system which consumes twice as much primary energy to deliver the same amount of useful heat. See Figure 6-2.

PRIMARY FUEL USAGE PER MILLION (10^6 Btu)

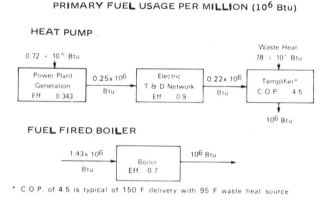

* C.O.P. of 4.5 is typical of 150 F delivery with 95 F waste heat source

**Figure 6-2. Comparison of Heat Pump Operation
with Boiler Operation**

The heat pump typically recovers the energy in waste heat in the temperature range of 50F to 160F. Such low-grade heat is characteristic of industrial cooling water or warm water effluent from plant processes. Other sources include cooling-tower water, water used for quench tanks, as well as for cooling arc welders, extruders, air compressors, furnaces, injection molding and heat treating equipment. Refrigeration and air conditioning condenser water is also an excellent source of waste heat, as are overhead vapors from distillation processes. Solar collectors, water-to-air heat pumps, geothermal resources and warm exhaust stacks are further additions to the seemingly endless list of waste heat sources that can be put to work.

Examples of waste heat pumping at several of the many installations of the high temperature heat pump, illustrate that their use has resulted in increased efficiency and a substantial savings in energy and costs.

TYPICAL APPLICATIONS

Make-Up Air Heating

Waste heat has been put to productive use at Robinson-Nugent, Inc., an electronics firm in New Albany, IN. The firm installed a carefully engineered system that saves money and energy by recycling and redirecting heat in its 48,000 square foot manufacturing facility.

All the normally wasted heat—from injection molding machines, workers, and the lighting system—is collected by several 20-ton water-to-air heat pumps. This heat pump system, in turn, heats and cools the entire 48,000 square foot manufacturing facility and the adjacent offices; and, in addition, stores excess heat in storage tanks and in an overhead water-loop piping system. Elsewhere, circulating water removes heat from oil coolers for the molding machines. This heat is also added to the water-loop. Refer to Figure 6-3.

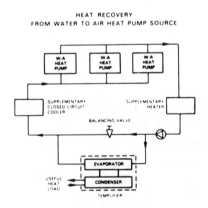

Figure 6-3

The heat stored in the water and storage loop heats the unoccupied plant on winter nights and during weekends. In the summer, the water-to-air heat pump system provides cool, comfortable air conditioning by removing unneeded factory heat.

The heating system is integrally tied to the pollution control system. So getting air, coming from the cold outside, hot enough to replace air removed by the pollution control system means really squeezing the most out of the plant's waste heat.

The Heat Pump System. A separate high temperature heat pump system was installed to heat the outside make-up air entering the plant.

The pollution control system captures toxic fumes from the plating department and whisks them at the rate of 20,000 cfm, into a scrubber.

The air from outside brought in to replace the air removed by the pollution control system has to be heated. This is done by make-up air heating coils using hot water from the industrial heat pump.

The industrial heat pump removes heat from the plant's water loop and heats glycol up to 150 degrees, which in turn heats the make-up air—even when the temperature dips to zero degrees, which has occurred several times since the plant first opened in 1977.

The heat pump also saves quite a bit over a more traditional resistance heating system. The heat pump's coefficient of performance—a measure of efficiency—is over three.

Hospital Laundry

The Crozer-Chester Medical Center on 67 acres in suburban Philadelphia has 25 buildings and has installed a new energy conservation system that will save countless thousands of dollars in the years ahead. Part of this system uses an industrial heat pump to transfer waste heat from the medical center air conditioning plant to the hot water used in the laundry. This alone is expected to save the hospital over half a million dollars in the next ten years.

In continual daily operation from 7 a.m. to 4 p.m., the laundry processes 5 million pounds of linens a year. There are five special "pass through" washers/extractors that handle 18,000 pounds of laundry a day. Soiled linen goes in one end; clean laundry is taken out the other end. The laundry uses more than 20,000 gallons of hot water a day. This has to be at 160F so that, when used in conjunction with detergents and bleach, it is hot enough to kill most germs or staph infections.

Before installing the waste heat recovery system, the laundry was the largest user of steam at the medical center. With the new system, the heat pump uses the waste heat in condenser water from the center's centrifugal air conditioning chiller. This waste heat ranges from 90 to 95F. The heat pump temperature amplifies the incoming water to 150F where it is further heated by steam to the 160F needed by the laundry. See Figure 6-4.

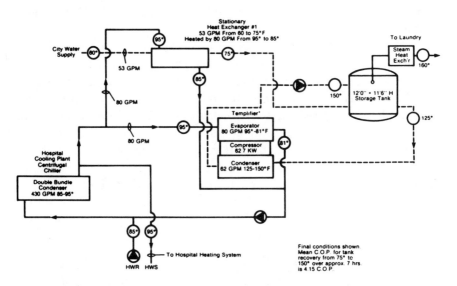

Figure 6-4. Crozer-Chester Hospital Heat Recovery System

Daily Cycle Operation. The daily cycle can be looked upon as if the 10,000-gallon storage tank is dumped two times per day. After being drawn down, the tank is filled with city water preheated to 75 degrees and the contents of the tank are gradually heated by circulating 62 gpm through the heat pump's condenser. Thus, for the initial 2½ hours after filling the entire tank, contents will be heated from 75 to 105F and progressively to 150F, requiring approximately 7 hours to recover from 75 to 150F.

Output of the heat pump varies with leaving condenser temperature. The average output per hour and coefficient of performance (COP) on applications of this sort are properly determined by using *mean* leaving hot water temperature (not the final temperature). Thus, the *mean* hot water temperature in this application is 135F. At that mean, heating capacity of the heat pump is 834,000 Btu per hour and the COP is 4.15.

Economic Evaluation. This system is saving a significant amount of money in reduced energy costs. The cost of oil to make steam has become very expensive and is going to become even more expensive in the future. In 1972, for example, the cost of the No. 6 oil used at Crozer-Chester was 6½ cents a gallon. Eight years later that same oil cost 87 cents a gallon.

A computerized economic evaluation was performed for the engineer to determine the projected cost savings. At the time of the study, electricity cost 4.86 cents per kilowatt-hour, and oil cost $5.72 per million Btu. General inflation was assumed to be 10 percent a year, while electricity was predicted to increase by 13 percent a year and oil by 17 percent a year.

With these design conditions, the heat pump system was expected to save $28,000 in reduced energy cost in its first full year of operation. This will increase to over $100,000 during the tenth year of operation. The cumulative net savings over the 10-year depreciation period, after deducting all maintenance cost and costs of ownership, will be $564,000 ($364,000 in terms of 1980 dollars).

The compounded annual rate of return on investment comes to 119 percent, with a simple payback period of less than one and a half years.

Space Heating

Another major commercial application is space heating in office and telephone company buildings. Typically, these structures have a high internal load, plus humidity control requirements where computer or electronic equipment is utilized.

The heat pump taps the waste heat generated by the computers or electronic equipment, gives it a boost in temperature and uses it for space heating, domestic hot water, and even for sidewalk snow removal system in more northern climates.

Telephone Building Installation. A telephone switching center in Pennsylvania is typical. It houses large office areas and each day it used to be filled with scores of operators who were needed to handle calls.

The building's function has now changed. Prior to the installation of the heat recovery system, a steam system was used for heating. Today one entire floor houses sophisticated electronic components and systems which are air conditioned year round to maintain constant temperature and humidity conditions.

The heat pump heat recovery system eliminates the use of steam by tapping the heat picked up by the 270-ton air conditioning system's chiller. See Figure 6-5.

The year-round cooling load supplies 90F warm water from the condenser to the heat pump. This source water would have normally been channeled to a cooling tower and the heat dissipated into the air. The heat pump amplifies this heat to produce hot water at 150F and passes it on to the subsystems directly to the coils in the air handlers for space heating and reheating, to radiators for perimeter space heating, to provide domestic hot water, and when needed, to the sidewalk snow melting system.

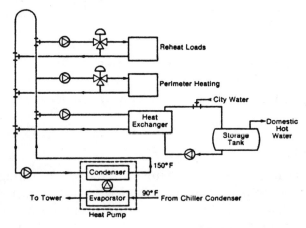

Figure 6-5. Waste Heat Recovery System at Bell Telephone Co.

Energy Savings Achieved. The new heat recovery system in this building has already had a dramatic effect. In November 1978—before the new system was installed—230,000 pounds of steam were used. This was cut in November 1979 to just 55,000 pounds of steam. This energy savings translated into dollars and cents will average over $27,000 per year over the 10-year analysis cycle.

As part of the justification process, a detailed economic analysis of the heat pump system was performed. At the time of the study, steam cost was $5.87 per 1,000 pounds. The estimated savings indicated the installation would pay for itself in about two years and produce almost a 48 percent return on the retrofit investment.

Industrial Process Heating

An industrial heat pump at PPG Industries' glass fabricating plant recovers normally wasted heat from process line air compressors. This recovered heat is used to provide hot water for parts rinsing and other process line applications. The energy recovery system significantly reduces the need for expensive natural gas and currently saves nearly $13,000 a year in reduced energy cost.

Installed in 1979, the industrial heat pump uses waste heat from large air compressors' cooling water. By extracting energy from this waste heat, it heats return water at 120F, boosts its temperature and sends it back to the plant, where it heats demineralized rinsing water, preheats wash and domestic water, and even funnels space heat to the plant maintenance area. In all, the system provides much of the plant's current hot water needs.

The Heat Recovery System. A closed-loop circulating water system is used to cool the plant's large air compressors. The warm water is piped to the heat pump's evaporator where the waste heat is removed, cooling the water to about 85F and returning it to the air compressors for reuse. A second process hot water loop feeds the plant's needs and delivers return water to the heat pump condenser at about 120F where heat is added and the temperature increased. This hot water returns to the plant to do its work.

The first priority use of the heated water is for the demineralized water needed for rinsing the glass panes before their edges are fused together. The water is demineralized to eliminate any spots that might otherwise occur. The water from the heat pump goes through a heat exchanger in the demineralized water loop to provide the 120F hot rinse water.

Since excess heat is still available, the heated water then proceeds to the plant's hot water distribution system. In this system it preheats city water entering at a cold 50F. This water is used for

Source: Templifier®

Figure 6-6. Heat Pump Installation at PPG Industries Glass Fabricating Plant

both domestic purposes and in the soft water system connected to the process glass washer. At the washer, the water is further heated by electric immersion heaters to the 160F needed for detergent action. By preheating in this manner, much less electric heat is needed from the immersion heaters.

But there is still more heat that has not been used. In the winter, this excess heat is sent to a fan-coil unit to provide space heat for the maintenance area. In the summer, the heat is simply vented outdoors by automatic duct dampers.

Economic Justification. When operation is increased to the full design load and coefficient of performance (COP) is 4.3, the heat pump will deliver approximately 4.3 Btu of energy to the plant (in place of natural gas) while using only 1 Btu of purchased electrical energy. The other 3.3 Btu is "free" energy recovered from the compressor cooling water.

At the time PPG engineers conducted an economic justification, the cost of electricity was 1.78 cents per kilowatt-hour and the cost of gas was $3.20 per thousand cubic feet.

At the planned part load operation initially, the annual net reduction in gas consumption came to 4150 mcf (or million Btu) a year. Subtracting the incremental cost of the electricity to operate the heat pump, and considering the savings in reduced operation of the electric immersion heaters, the net savings were $13,900 a year. At this rate, the heat pump system (Figure 6-6) will pay for itself in less than 3 years.

When the plant expands and the system operates at full design load, two years' energy cost savings will about equal the heat pump system's installed cost.

REFERENCE

1.   R. C. Niess, "High Temperature Heat Pump Applications," *Advances in Energy Utilization Technology,* Proceedings of the 4th World Energy Engineering Congress. The Fairmont Press, Inc., Atlanta, GA 30324.

7

Condensation Heat Recovery
—Direct Contact

Condensation Heat Recovery is defined as the process of cooling exhaust gases below their dew point and recovering greater than 50% of the latent heat content of the exhaust gas as well as a significant amount of sensible heat. This recovery process has two distinct features: the amount of heat recovered is much greater than with standard technology items such as economizers, etc. (thermal efficiency improvements of 10 to 15% are typical), and the recovered heat is in the form of a hot water stream ranging from 110F to 180F depending on the type of equipment.

In this chapter the basic operating conditions for condensation heat recovery and description of direct contact heat exchangers will be provided.

BACKGROUND

While Condensation Heat Recovery is relatively new to the U.S. market it has been applied for ten years in Europe. Early studies were conducted by Gaz de France and L'Industrial de Chauffage in France in 1971. Systems were installed as early as 1972 and have been operating for nine years. A German company, Froling, entered this market in 1975 with a unit designed primarily for small commercial applications and have over 600 units installed. Flakt Fabrieken, a Swedish firm, has been marketing a unit designed for heat recovery and pollution control for black liquor recovery boilers in the pulp and paper industry and have over 50 units installed throughout Europe, some of which have been in operation since 1974.

New entries to the market include an English company, Bayless-Kenton (installed 17 units since 1979), Heat Extractor Corporation (installed over 100 units since 1976), and A. O. Smith with one unit installed in 1981.

With a total of approximately 800 units in operation, this technology has been established and documented test results indicate a typical thermal efficiency improvement of 10–15% when firing natural gas with a minimum of maintenance effort.

The basic principle behind this heat recovery technique is the direct contact heat transfer between the exhaust gases and a cold water stream. This type of heat transfer has been applied in the chemical and petroleum industries for many years in areas such as reactor gas quenching, barometric condensers, feed gas desuper-heaters, etc. Essentially, the extension of this heat recovery technique to the field of exhaust gas heat recovery is due to the greater value of the recovered heat and has not required any new research effort, but primarily development efforts to economically apply this heat exchange technique to a new application area, namely exhaust gas heat recovery.

EQUIPMENT DESIGN

This discussion will present the basic features of a Direct Contact Condensation Heat Recovery system. This will be a generic discussion and the reader should note that manufacturers have provided various configurations in their equipment design to optimize the recovery process. The configuration discussed below is merely the basic approach and does not constitute a recommended design.

A direct contact heat exchanger is a vessel where the process stream and cooling fluid are commingled directly. There is no heat flow resistance from a separating wall, nor complexity due to arranging this wall for maximizing the heat transfer surface. The attractiveness stems from the potential economy and simplicity, plus the ability to handle a wide variety of fluids and solids—often under conditions that would cause excessive fouling, corrosion or thermal stressing of conventional tubular equipment.

The four general classifications of direct contact gas liquid heat transfer are:

1. simple gas cooling
2. gas cooling with vaporization of coolant
3. gas cooling with partial condensation
4. gas cooling with total condensation

In the application of flue gas condensation heat recovery, we are concerned with items 2 and 3 listed above. The following is a description of the processes occurring within the heat exchange tower.

Figure 7-1 shows the basic Direct Contact Condensation Heat Recovery system. The exhaust gas enters the bottom of the tower and is cooled by direct contact with a cold water stream entering the top of the unit. The exhaust gas is generally cooled to a temperature in the range of 100–110F and the cold water stream heated to 120–130F, depending on the type of tower internals and inlet gas temperature and humidity.

The heated water stream is passed through a secondary heat exchanger where the recovered heat is transferred to a clean water stream. The use of a secondary exchanger is generally recommended to isolate the water in contact with the exhaust gas (referred to as the working fluid) from the rest of the heat distribution system. The working fluid absorbs a small amount of nitrogen oxides from the exhaust gas which results in a liquid pH of approximately 5. As many users of the equipment are concerned about the potential problems from circulating slightly acidic water through their heat exchangers or boilers (in the case where the water is used for boiler make-up), the use of a secondary exchanger has become standard.

The rejection of the recovered heat is critical to the proper operation of the heat recovery unit. The exhaust gas outlet temperature is directly dependent on the inlet water temperature to the top of the tower, and as the water temperature increases, the outlet gas temperature increases, which results in decreased heat recovery rates. Consequently, utilization of the recovered heat is a high priority, and every effort should be made to maintain as low a water inlet

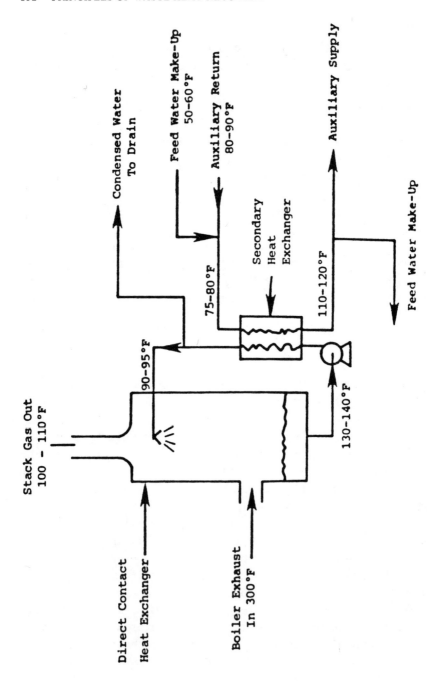

Figure 7-1. Condensation Heat Recovery Process

temperature as feasible. Methods for using the recovered heat and maintaining low water return temperatures will be discussed below.

While the inlet water temperature determines the outlet exhaust gas temperature and consequently the thermal recovery rate, the temperature of the working fluid to the secondary heat exchanger determines the ease of utilization of the recovered heat. Typical uses of the recovered heat will be discussed in the section on Applications; however, it is quite obvious that the higher the working fluid temperature the wider the range and the greater the ease of utilization.

The factors that limit the working fluid temperature are determined by the direct contact nature of the process. The total heat recovery process can be divided into four separate sections as shown in Figure 7-2. Sections 1 and 4 are inlet and outlet sections required for exhaust gas inlet, water storage at the bottom of the tower to provide a liquid head for the working fluid pump, inlet water distribution piping and outlet exhaust gas—water disengagement space.

In Section 2—Humidification, the exhaust gas contacts the water stream and the process of adiabatic humidification occurs. The exhaust gas is cooled to its adiabatic saturation temperature by vaporizing the working fluid until the gas is saturated. As the liquid is not heated during this process since all of the heat released from the gas is used to vaporize, the adiabatic saturation temperature represents the maximum limitation on the temperature of the heated working fluid. The actual working fluid temperature will be determined by the type of tower internals used to promote heat and mass transfer and the volume of contact space provided.

Adiabatic Saturation Temperature

The adiabatic saturation temperature of the exhaust gas is determined by its inlet temperature and humidity according to the equation

$$H_S - H = c_S \ (t - t_S)/\lambda \qquad\qquad 7\text{-}1$$

where H_S = adiabatic saturation humidity, lb of water per lb dry gas
H = inlet gas humidity, lb of water per lb of dry gas
c_S = humid heat of gas = .24 + .45 H, Btu/lb—$^\circ$F

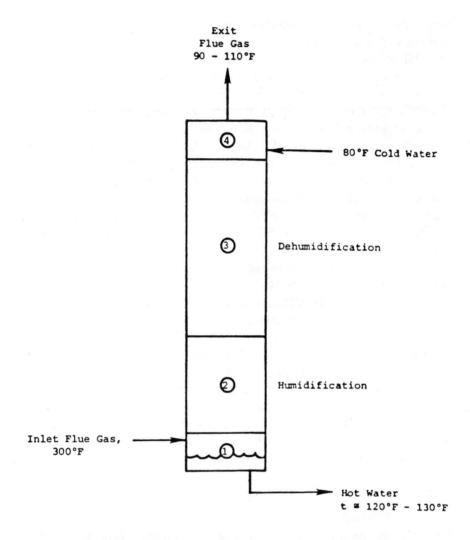

Figure 7-2. Direct Contact Heat Exchanger Tower Conditions

t_S = adiabatic saturation temperature, °F
t = inlet gas temperature, °F
λ = latent heat of vaporization, Btu/lb

Figure 7-3 shows the relationship between the adiabatic saturation temperature and inlet gas temperature and humidity. The conditions for a typical natural gas or oil fired boiler are shown as well as exhaust streams with higher humidities typical of dryer applications.

The actual working fluid outlet temperature approach to this maximum will vary from 5 to 30F depending on the tower design.

In the dehumidification section (3), the cold water provides the driving force to condense part of the water from the flue gas recovering both the latent heat added to the flue gas in Section 2 and part of the additional latent heat of condensation associated with the water vapor in the inlet flue gas. The flue gas exits Section 3 with less water content than it entered Section 1, resulting in a net recovery of latent heat. The amount of water in the flue gas exiting Section 3 is determined by the vapor pressure of the water at the outlet temperature.

Most direct contact heat exchangers use the following contacting methods:

1. Baffle-tray towers
2. Spray chambers
3. Packed towers
4. Cross-flow tray towers
5. Pipeline contactors

Generally, only the first three items mentioned above are used for heat transfer service. Cross-flow tray towers (item 4) are too expensive except for vacuum fractionation applications where low pressure drop and heat and mass transfer considerations are critical. Pipeline contactors (item 5) are not considered suitable for flue gas condensation heat recovery systems because of their high pressure drop. Therefore, the following discussion is limited to the first three items mentioned above; namely, baffle-tray towers, spray chambers, and packed towers.

It is important to realize that in this type of equipment there is basically a tradeoff between heat transfer performance and gas side pressure drop. Table 7-1 shows the relative heat transfer and pressure drop characteristics of the three devices. The spray chambers have low heat transfer performance per unit volume and correspondingly

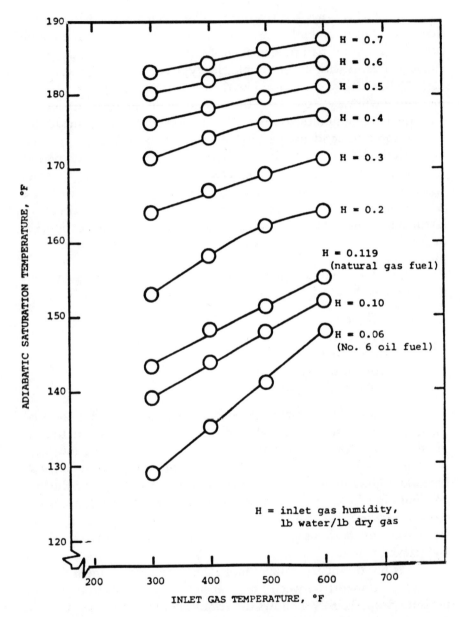

Figure 7-3. Adiabatic Saturation Temperature Dependence on
Inlet Gas Temperature and Humidity

Table 7-1. Heat Transfer/Pressure Drop Characteristics

Type	Spray Chamber	Baffle-Tray	Packed Tower
Heat Transfer Performance/ Unit Volume	Low	Medium	High
Pressure Drop/ Unit Volume, Inches of Water	Low <0.1	Medium 0.5–0.8	High 5–10
Hot Water Temperature, °F	Low 100–110	Medium 115–125	High 130–140

low pressure drop requirements ($<$ 0.1 inches of water). For a majority of the small industrial boiler applications, these units have been placed on forced-draft or balanced draft boilers without requiring any additional fan capacity with associated controls and increased operating costs.

The baffle-tray tower represents a compromise between heat transfer characteristics and pressure drop. The heat transfer performance of a baffle-tray tower is superior to a spray chamber device, but it has a correspondingly greater pressure drop requirement. It would be anticipated that in some applications, additional fan capacity could be required; however, as the pressure drop is normally limited to 0.5 to 0.8 inches of water, there are many situations that do not require a fan and controls.

The packed tower has the highest heat transfer performance by virtue of the large amount of interfacial area on the packing surface; but it also has the highest pressure drop requirement (5–10 inches of water) due to the limited flow area for the gas.

The decision as to which type of heat transfer-pressure drop contacting device to use must be based on the individual requirements such as hot water temperature required, existing fan capacity, available floor space, retrofit (breeching and load exchangers) requirements, etc.

THERMODYNAMICS AND
EFFICIENCY INCREASE

The efficiency improvement from the application of Direct Contact Condensation Heat Recovery can be easily visualized by referring to the Heat Recovery curve, Figure 7-4. Cooling the flue gas from 300F to the dew point of 140F results in an efficiency increase of 3 percent. Further cooling, resulting in condensation of water vapor, increases the rate of heat recovery dramatically and a 10 percent total efficiency improvement is obtained at a 100F outlet temperature to the stack. This indicates the relative importance of achieving the condensation resulting from the low stack gas temperature.

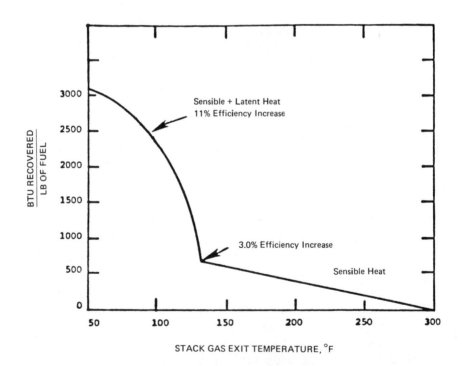

Figure 7-4. Heat Recovery Curve (Natural Gas Fired Boiler)

The boiler efficiency improvement resulting from the installation of a condensation heat recovery (CHR) unit is determined by:

1. Fuel type

2. Boiler flue gas exit temperature (or direct contact CHR inlet temperature)

3. Heat sink temperature

4. Low level heat requirement

5. Fuel moisture content

6. Combustion air humidity

The first four items—fuel type, flue gas temperature, heat sink temperature, and low level heat requirement—are the most critical, with fuel moisture content and combustion air humidity having a secondary effect for natural gas-, oil- or coal-fired boilers. Applications using lignite or biomass fuels with high moisture content should be investigated with the equipment manufacturers. The effect of combustion air humidity is such that in a high humidity area the resulting efficiency increase can be 1 percent higher than predicted.

The DCCHR/flue gas temperature and fuel type determine the basic efficiency improvement (ΔE_b). This correlation is presented in Figure 7-5 for an outlet stack gas temperature of 100F. Variation from this outlet flue gas temperature from the heat recovery unit can be accounted for using the factor F according to the formula

$$\Delta E = F_1 \times \Delta E_b \qquad\qquad 7\text{-}2$$

where ΔE is the total efficiency improvement.

Figure 7-6 shows the relationship between F_1 and varying outlet stack temperatures. It is then necessary to determine the outlet stack temperatures from the unit. This is fixed by the temperature of the water into the top of the unit and the performance of the direct contact exchanger. The inlet water temperature is fixed by the cold water temperature into the secondary heat exchanger (see Figure 7-1) and the performance of the secondary heat exchanger. Generally, a 20F temperature differential between the cold water temperature and the stack gas temperature should be sufficiently accurate for a preliminary evaluation.

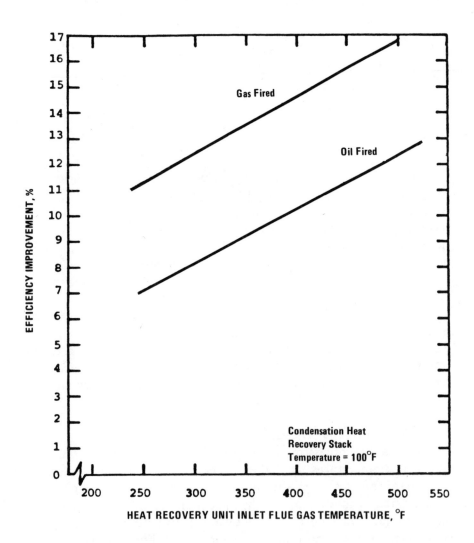

Figure 7-5. Efficiency Variation with Heat Recovery Unit
Inlet Flue Gas Temperature

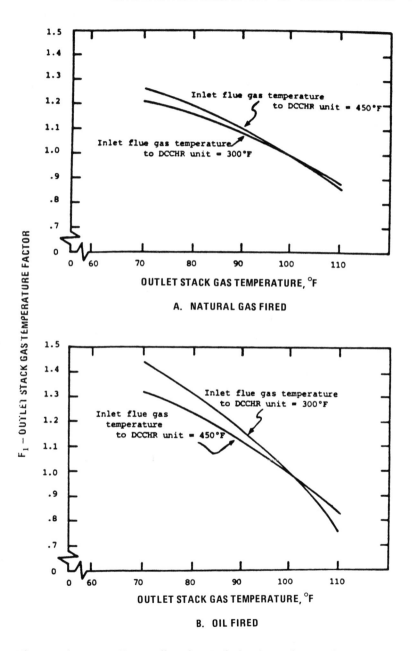

Figure 7-6. Factor F_1 to Allow for Variation in Outlet Stack Temperature

The following examples illustrate the use of Figures 7-5 and 7-6:

(1) Fuel   Natural Gas
    Inlet Stack Temperature   450F
    Inlet Cold Water   60F
    Outlet Stack Temperature   =   60F + 20F = 80F
    ΔE_b =   14.5% (Figure 7-5)
    F_1 =   1.19 (Figure 7-6)
    $\Delta E = 14.5 \times 1.19$ =   17.3%

(2) Fuel   Oil
    Inlet Stack Temperature   300F
    Inlet Cold Water   70F
    Outlet Stack Temperature   =   70F + 20F = 90F
    ΔE_b =   7.2% (Figure 7-5)
    F_1 =   1.18 (Figure 7-6)
    $\Delta E = 7.2 \times 1.18$ =   8.5%

It should be noted that although the efficiency improvement percent (ΔE) for oil-fired boilers is generally less than for natural gas-fired boilers (due to the lower fuel hydrogen content in oil), the present costs of these fuels is such that the dollar value of the recovered heat for oil-fired boilers may be greater than for natural gas boilers.

To further illustrate the impact of condensation heat recovery on the operating efficiency of a boiler, the effect on the ASME heat loss is evaluated. Example results for three cases are presented in Table 7-2.

The output from the ASME heat loss calculation consists of the following

- Dry gas heat loss, Btu/lb or % of input (also referred to as sensible heat loss)

- Latent heat loss, Btu/lb or % of input — consists of latent heat loss associated with water content of fuel and latent heat loss from combustion of H_2 in fuel

- Radiation heat loss, Btu/lb or % of input (loss associated with losses through the external boiler envelope)

Table 7-2. ASME Heat Loss and Efficiency Improvement

Case No. Type of Unit Fuel	1 Boiler Natural Gas	2 Boiler No. 6 Oil	3 Boiler No. 6 Oil
Fuel Characteristics			
C%	75	86	86
H%	25	13	13
S%	0	1	1
O%	0	0	0
N%	0	0	0
Water %	0	0	0
Higher Heating Value, Btu/lb	22,500	19,500	19,500
Combustion Air Temperature, $^{\circ}$F	80	80	80
Relative Humidity, %	50	50	50
Capacity, MMBtu/hr	50	50	20
Loading Rate, %	100	100	100
Flue Gas Conditions			
Temperature, $^{\circ}$F	310	320	450
Stack O_2, %	2	3	4
Humidity, lb water/lb dry gas	.125		
ASME Heat Losses			
Dry Gas Heat Loss, %	4.8	4.6	7.6
Latent Heat Loss, %	11.5	6.9	7.3
Other Losses	2.5	2.5	3.0
(radiation and miscellaneous are unaffected and considered constant)			
Total Losses, %	18.3	13.5	17.9
Efficiency, %	81.1	86.4	82.1

Condensation Heat Recovery Unit

Exit Gas Temperature, $^{\circ}$F	100	100	100
Relative Humidity, %	100	100	100
Heat Recovery			
Sensible Heat, %	3.9	4.2	7.2
Latent Heat, %	8.7	4.0	4.0
Total, %	12.6	8.2	11.2
Net Losses			
Sensible, %	0.4	0.4	0.4
Latent, %	2.8	2.9	3.3
Other, %	2.5	2.5	3.0
Total Losses, %	5.7	5.8	6.7
Efficiency, %	94.3	94.2	93.3

- Miscellaneous losses, Btu/lb or % of input (generally determined by boiler manufacturer)

Condensation heat recovery results in recovery of both sensible heat associated with the dry gas losses and latent heat associated with the moisture content of the flue gas.

For the example Case 1 shown in Table 7-2, a gas-fired boiler operating with a 310F stack temperature, the ASME sensible and latent heat losses are 4.8% and 11.5%, respectively. Reducing the exit gas temperature to 100F results in recovery of 3.9% of the sensible heat loss and 8.7% of the latent heat loss for a total heat recovery rate of 12.6%. The operating efficiency of the boiler is increased from 81.1% to 94.3%. This efficiency increase can be used to either decrease the fuel consumption for a given operating rate or allow an increase in the heat output from the boiler plant with no increase in fuel consumption or *emissions.* The increase in output with no increase in emissions can be particularly effective in EPA nonattainment areas where the plant output is limited by the emissions produced by the boiler.

Case 2, Table 7-2, shows a similar analysis for an oil-fired boiler operating with a 320F stack temperature. The ASME heat losses for sensible and latent heat are 4.6% and 6.9%, respectively. Reducing the exit gas temperature to 100F results in recovery of 4.2% of the sensible heat loss and 4.0% of the latent heat loss for a total heat recovery rate of 8.2%. The operating efficiency of the boiler is increased from 86.4% to 94.6%. As in the gas-fired application, the higher efficiency can be used to either decrease the fuel consumption for the same heat output from the boiler plant or allow an increase in the boiler plant capacity with no increase in fuel consumption and a *decrease in emissions!*

The emissions reduction results from the scrubbing nature of the direct contact process which removes particulate and SO_x from the flue gas stream. The magnitude of the emissions reduction has not been quantified; however, qualitatively the removal of particulate and SO_x are known to occur. If the emission reduction potential of the heat recovery process is of particular interest a qualified consultant or the equipment manufacturer should be contacted to evaluate the potential performance of this equipment.

The previous two cases were examples for boilers operating with relatively low stack temperatures approaching the minimum acceptable with standard technology. The heat recovery rates of 12.6% for natural gas-fired and 8.2% for oil-fired application represent the minimum recovery rates. Case 3 shows an example for an oil-fired boiler operating with a 450F stack temperature to illustrate the higher potential for recovery rates. The sensible heat loss without condensation heat recovery is 7.6% and the latent heat loss is 7.3%. The heat recovery rates with condensation heat recovery are 7.2% for the sensible heat and 4.0% for the latent heat for a total recovery rate of 11.2%. The operating efficiency for the boiler increases from 82.1% to 93.3%.

OPERATION AND MAINTENANCE

1. *Circulating Water Flow.* The circulating water flow rate is critical to the proper operation of a DCCHR unit. The ratio of liquid to gas flow rates (L/G) is a major factor in determining the heat transfer performance of the equipment. Typical L/G ratios range from 3–10 with the lower range required for the more efficient packed tower design and the higher range being necessary for spray-type tower operation. The baffle-tray design falls midway between the two at approximately 7. The pump operating costs are generally around 1 percent of the value of the recovered heat.

2. *Fan Operation and Control.* If a fan is required its operation and control are critical to proper boiler control. Control procedures used have been quite varied, including a motorized damper system controlled by a pressure sensor at the boiler flue gas outlet, variable speed fans controlled by the fuel input to the boiler, and constant flow rate systems with a fixed flue gas flow rate. The fixed flow rate design either draws an increased air flow through the boiler increasing the excess air operation at low loads or draws the required air down the original stack to mix with the hot flue gas flow at low loads. All of the above methods have proven satisfactory.

The use of a fan on the system requires control of the start-up procedure to prevent flame blow-off from excessive draft in the fire box. Generally, opening the damper controlling the flue gas flow to the DCCHR is delayed during start-up until the flue gas reaches its design temperature.

Fan operating costs are generally 2–4 percent of the value of the recovered heat.

Systems requiring a fan are considerably more complicated than a simple tower system. This is a distinctly negative feature in commercial or small industrial applications where simplicity is required. Larger industrial installations with sophisticated controls and greater operator competence are less affected by the fan requirement.

3. *Water Treatment.* Gas-fired applications have had no water treatment problems. The pH of the working fluid is generally between 4 to 5 and dilution of the condensate overflow with normal plant water effluent is generally considered sufficient to meet EPA regulations.

Oil-fired applications require some water treatment to neutralize the H_2SO_4 in the condensate overflow. This has been accomplished by either piping the overflow to existing water treatment facilities or, in other cases, a simple caustic neutralization with pH control of the effluent at 7–8 has been used.

Generally, operations and maintenance requirements for DCCHR equipment are minimal if the system is properly designed.

Other Factors Affecting Boiler and Fuel Efficiency

1. *Excess O_2.* The retrofit of a DCCHR unit to a boiler results in a significant decrease in the impact of excess O_2 levels on boiler efficiency. Without DCCHR an increase in the excess O_2 level of 1 percent decreases boiler efficiency by 0.36 percent. With DCCHR the energy loss associated with the flue gas is decreased and a 1 percent increase in excess O_2 only decreases the boiler-DCCHR system efficiency by 0.24 percent. This effect is shown in Figure 7-7. This results in the overall efficiency being less dependent on excess O_2 levels and can affect the economic impact of sophisticated excess O_2 controls.

2. *Standby Losses.* All boilers have thermal losses associated with start-up and shutdown. This effect can be minimal in baseload boilers operating 24 hours/day or can be quite significant in swing

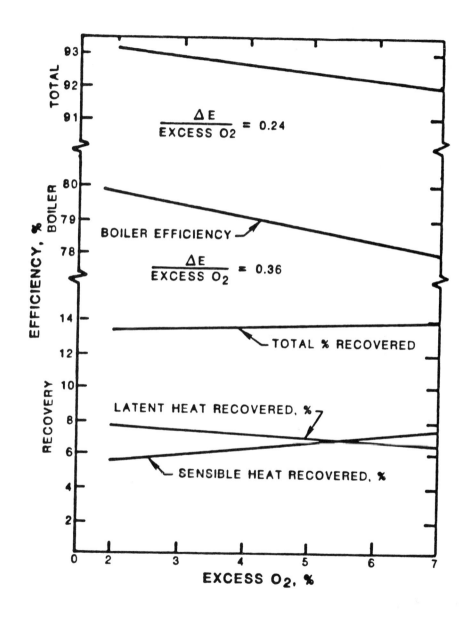

Figure 7-7. Variation of Efficiency with Excess O_2

boilers operating with frequent load variations and idle periods. The retrofit of a DCCHR greatly decreases the thermal losses from boiler draft during idle time by decreasing the stack temperature and providing greater resistance to cold air flow through the boiler. The net result is an increase in the boiler's yearly fuel efficiency that can be as high as 5 percent.

The standby losses can also be eliminated by virtue of the increased capacity of the baseload boiler. If the swing boiler has been operating at levels of 10–15 percent of the baseload capacity, a high percentage of the swing cycles can be eliminated resulting in a significant savings of the energy associated with the start-up and shutdown cycle.

Emissions Impact

This process has the unique feature of resulting in both energy recovery and emissions reduction. Minimally, the plant emissions are reduced by virtue of the fact that the boiler efficiency is increased by roughly 10 to 15 percent. In addition, when burning natural gas, a small percentage of the NO_x emissions are reduced by condensation of nitric oxide. In sulfur-bearing fuel applications, the SO_2 emissions can be reduced significantly by the use of a water bath that is highly buffered, resulting in an alkaline water spray. The resulting water bath must be discharged in a pH range of 6–8 which requires additional caustic storage, metering pump, and pH control or inclusion of this wastewater stream in the plant's existing water treatment facility. The emission reduction potential of this equipment can have significant effect in areas that the EPA had judged to be nonattainment and could result in an increased plant capacity combined with an emissions reduction. It should be pointed out that the SO_x emission reduction from the scrubbing tests should be provided for in any installation that is dependent on the expected emission reduction.

EQUIPMENT COSTS

As this technology is in its early commercialization stage, the costs presented must be considered as only a rough estimate. However, the cost figures presented are sufficiently accurate for a preliminary evaluation of a potential application.

The capital costs vary considerably within the industry. Some companies are very new to the market while others are considered old and established after five years of operating experience. The cost variations are also due to considerable differences in design approach and equipment construction. The commercial units also differ with respect to heat transfer performance and temperature level of the hot water produced which can significantly affect the payback period for certain applications. The cost figures presented below represent the only available data at this time. These figures could be revised considerably due to a more effective design approach, decreased manufacturing and engineering costs due to increased sales volume and lower costs due to increased competition.

Figure 7-8 shows the range for equipment and installed costs. The installed cost can be as much as three times the uninstalled capital equipment cost due to retrofit difficulties. The cost items making up a complete system are:

- condensation heat recovery exchanger
- secondary exchanger
- pumps
- fans
- flow controls
- instrumentation
- stack
- retrofit exchangers
- installation

The following cost estimating equation can be used in a preliminary evaluating procedure:

$$\text{Equipment} = \$10,000 \left(\frac{\text{Boiler Size, } 10^6 \text{ Btu/hr*}}{3.348} \right) 0.7$$

The solution of this equation is shown as the dashed line in Figure 7-8. An installation difficulty factor between 2 and 3 should be used depending on the estimated difficulty for retrofit.

---

*Or $\sim 10^3$ lb steam/hour

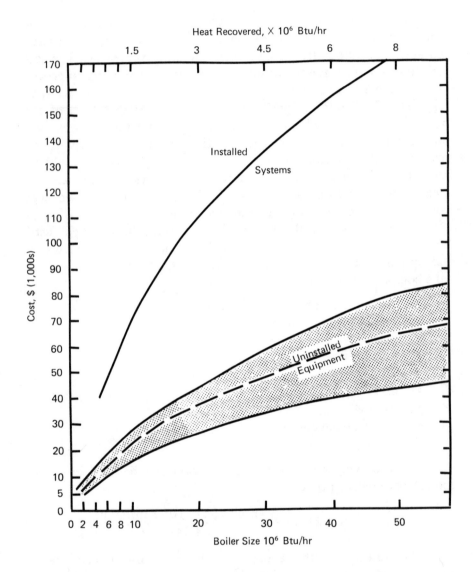

**Figure 7-8. Uninstalled Equipment and Installed System Costs
for Condensate Heat Recovery versus System Size**

Operating costs for these units come from fan and pump horse-power requirements. These generally range from 5 to 10 percent of the value of the recovered heat. The 5 percent figure would apply for an application with limited distribution of the hot water (i.e., boiler house use only) and a low pressure drop condensation heat recovery unit. The 10 percent figure would apply to a system distributing hot water a distance of 100 feet and/or a higher pressure drop heat recovery unit.

8

Recovering

Process Waste Heat

Acknowledgement: This chapter is based on Manual 8, Industrial Energy Conservation Manual, authored by William F. Kenney and copyrighted by Massachusetts Institute of Technology.

SOURCE OF WASTE HEAT
COMBUSTION FLUE GASES

When fuel was cheap, fired-process equipment was designed with relatively high stack temperature and excess-air rates. With today's rising fuel prices, much of this equipment in use represents a large potential for recovery of waste energy from flue gases.

Both the flue-gas temperature and the available draft are important in the economics of projects to improve fired-heater efficiencies. Present designs would have stack temperatures of about 350°F for sulfur-containing fuels and as low as 250°F for "sweet" (sulfur-free) fuels. However, reducing stack temperature greatly reduces the available draft from the stack, since

Theoretical draft (in. H_2O)

$$= \frac{12 \text{ X SH X (Density of cold air} - \text{Density of hot flue gas)}}{\text{Density of } H_2O}$$

(where SH is stack height, in feet). For existing equipment this is often partially compensated by operating at 10 percent excess air with a stack designed for 30–100 percent.

Draft becomes a major determinant in the economics of projects to improve the efficiency of existing heaters. To recover more heat from the flue gas, more tubes must be added in the convection section. This adds pressure drop and increases the draft required, while the available draft is being lowered. This conflict leads quickly to the first economic limit: the available natural draft (see Table 8-1). For small heaters, recovery is justified only up to the point where natural draft is available.

Providing for increased draft involves a step change in cost. An induced-draft fan or a new stack with associated ducting can generally be justified only for large (> 100 million Btu/hr) installations.

In a number of installations, induced draft has already been provided for one reason or another. Here the designer does not face step changes in costs to increase draft. Changing a fan rotor, a motor, or both is much less costly. In such systems it pays to achieve greater energy recovery.

A gas turbine is a common example of a forced-draft system. The exhaust from the turbine proper is generally in the range of 800–900°F, contains about 17 percent oxygen, and is at a positive pressure of a few inches of water. Turbine vendors have made extensive studies of waste-heat boilers and recuperators for their machines. In combined-cycle utility plants the exhaust is often used as preheated air for a conventional boiler. The key element in handling gas-turbine exhaust is pressure drop. Significant changes in driver performance occur if exhaust pressure exceeds the vendor's limits. The engineer must work closely with the vendor in proposing to recover energy from a gas turbine.

Heat transfer in flue-gas systems is expensive. Overall coefficients are low, generally less than 10 Btu/hr-ft^2-°F, and extended surface is the rule. The temptation exists to minimize capital costs by using the maximum temperature difference between the flue gas and the heat-recovery fluid. Before taking this route, the designer must carefully analyze long-term prospects and overall site utility balances, because large differences in temperature mean large increases in entropy and therefore low thermodynamic efficiencies. To the extent possible, a high-temperature heat source should be used to fill a high-temperature need.

A final consideration is that flue gas is dirty. Depending on the fuel, convection soot blowers and studded rather than finned tubes may be required. In dryers or calciners, various corrosive vapors or solids (other than the usual sulfur oxides) may be present.

Table 8-1. Effect of flue-gas temperature on stack draft.

Flue-Gas Temperature (°F)	Stack Height for 0.8-Inch Draft (ft)[a]
300	240
400	185
500	160
600	143
700	130

a. Includes stack friction and exit loss of 0.2 inch H_2O at bottom of stack.

REACTOR COOLING

Many reactions are exothermic and provide opportunities for heat recovery. However, more complexities are associated with recovery of this energy than with other opportunities. Typical problems that affect either the technical or the economic feasibility of any scheme include materials of construction, impact on production rate, fouling characteristics, pressure requirements, process control, safety, consequences of leaks into or out of the reactor, and geometry of reactor.

Reactors that require high process-side pressures are typically difficult to improve. In tubular LDPE reactors, a mile or more of double-pipe exchangers operating at 50,000 psi serves as a reactor. Product quality and production rate depend very much on the temperature profile of the reactants. Cooling water is generally used in the annular space to remove heat and control reaction temperatures. If this water were to be used to generate steam, several problems would be created. Because the temperature differences between reactants and coolant would be changed, the temperature profile of the reactants and hence the product capacity and quality would change, and the capability to deal safely with reaction runaway (mass and temperature of coolant) would be restricted. Of course, the shell-side geometry would also have to be changed. In such specialized reactor systems there is often little that can be done on a retrofit basis.

In less complex cases, simple approaches that preserve the essentials of the original design are often practical. In the somewhat rare case where additional boiler feedwater preheat cannot be provided more economically elsewhere, the substitution of boiler feedwater for cooling water is often straightforward and economical. In other cases feed can be preheated directly in the reactor or in its cooling channels. These approaches avoid additional problems in construction materials and phase changes, and involve minimum capital investment.

PROCESS COOLERS AND CONDENSERS

Condensers and coolers often provide the easiest access to recovery of waste energy. In a typical petrochemical plant a large share of the energy input downstream of the reactors is rejected in the condensers of a fractionation train, in an extraction solvent or absorbent recovery tower, or in product coolers. Viewed as energy sources, these streams have a number of advantages. They are plentiful, present fewer process complications, involve clean fluids, permit good design for heat transfer, have adaptable layouts, and create fewer materials problems. There are some disadvantages, however. In many processes temperatures are low, and thus the recovered energy is valueless. In others, tower pressures can be raised to increase the potential of the recovered heat. Then, however, relative volatility may be reduced, increasing the requirement for energy. Heating boiler feedwater or producing low-pressure steam is usually possible, and for large installations (50–100 million Btu/hr) heat pumps, Rankine-cycle power generation, or other machinery-aided recovery techniques may be economical.

HIGH-PRESSURE FLUIDS

Work can be recovered from some waste streams. In general, each Btu of work recovered corresponds to about 3 Btu of fuel saved.

Gas expanders are practical in a number of applications. These range from rather small streams at medium to high pressure in refrigerated systems to very large streams at relatively low pressures. For example, the overhead from the demethanizer tower in an ethylene plant is very cold at 400–500 psia. It is now relatively standard in new designs to use a turbine to let down the pressure of this stream for power recovery, but sometimes only for some additional refrigeration credits. These credits are reflected either in refrigeration-compressor horsepower savings or in improved ethylene recovery. There are multiple opportunities for this method of saving energy, and retrofits are often justified.

At the other end of the scale, flue-gas expanders for catalytic crackers are often practical. The gas stream here is at low pressure ($<$ 30 psig) and contains some catalyst fines and other foulants, but has the redeeming qualities of being very large and hot ($\sim$ 1,200°F). At present there are 13 axial-flow single-stage expanders in commercial service on the outlet of the regenerators in fluid catalytic crackers. Ten of these units were supplied by Ingersoll-Rand and three by the Elliot Company.

In summary, gas streams that are currently let down to lower pressures by throttling—and this includes steam—are good potential sources of work recovery. Recovery possibilities must be either directly related to the source of horsepower or totally independent (power generation or air compression) to avoid service-factor complications.

High-pressure liquids offer less horsepower recovery potential than gas streams. Just as it takes less work to pump a pound of water to a higher pressure than to compress a pound of steam through the same pressure difference, so also less energy is recovered in the reverse process. Only the large liquid streams offer practical opportunities.

In process plants, the various acid-gas (CO_2, H_2S) scrubbing processes require large volumes of absorbents to be cycled from high to low pressures. The pump often has a dual driver: a motor and a hydraulic turbine (Figure 8-1). Once flow has been established by the motor, return flow from the high-pressure tower drives the turbine, unloading the motor. The configuration shown in the figure includes a bypass, which is used when the energy-recovery path is unavailable or temporarily inoperable.

Mixed-phase applications face machinery-development problems. Efficient machinery to recover work from multiphase and flashing streams is not commercially available. Evolution of small quantities of gas, such as in the acid-gas scrubbing application, are tolerable, but with extensive flashing either the machinery will be damaged or energy of the vapor will go unrecovered. The replacement of throttle valves in refrigeration systems does not appear to be economic because of extensive flashing. The pressure at the outlet of LDPE reactors drops from 50,000 psi to several hundred, but energy recovery is not practical because of the several phases present and the unknowns about the thermodynamic properties of the inlet stream.

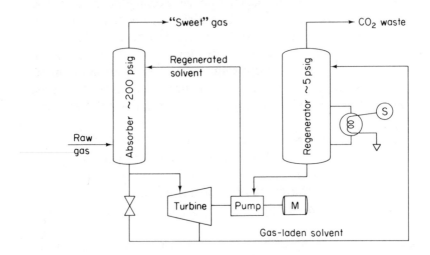

Figure 8-1 Liquid power recovery.

WASTE COMBUSTIBLES

Waste and vent gases have been recovered for fuel for some time. They are generally collected and burned in a dedicated furnace, boiler, or incinerator. Sometimes certain burners are dedicated for waste-gas use. Some companies add a nozzle to their conventional burners to distribute the waste gas and minimize control problems. In any system, firing controls must be able to compensate for variation in waste-gas heating value and flow rate without upsetting the process.

In some plants, mixing of recovered waste fuels into the plant fuel system may result in significant potential hazards, because the composition of the entire fuel system may be upset, or pipes may be corroded, or flashback characteristics may be altered or flow instabilities may occur, or export fuel may be contaminated. Corrosion may occur because of corrosive gases in the waste streams, and flow instabilities may be created by pressure fluctuations.

A refinery once managed to get some acid gases into its LPG product. Since the product was odorized and met other specifications, there seemed to be no problem until some retail customers started to complain that the bottoms of their cooking utensils were falling out!

Flare and vent-gas recovery systems are now becoming economically feasible. For both environmental and economic reasons, venting and flaring of waste hydrocarbons, even during upset conditions, is being curtailed. For all but major upsets, companies are finding it economical to recompress flare-line gas to the waste-fuel header. This requires a good estimate of the continuous flow and a flexible system. Machinery, control, and corrosion problems must be solved, on the basis of a good understanding of the flow variation and likely composition range of the gases to be recovered.

Recovery of flare-line gases poses safety problems as well. Often these gases will contain oxygen as well as the flammable "cats and dogs." Even mild compression of such a stream may provide enough heat to set off unsafe reactions. For example, olefine and acetylenes decompose in the pure state when heated. The presence of small amounts of oxygen can lower the temperatures at which these reactions occur, even though flammable limits are not approached.

Solid wastes can sometimes be recovered as fuel. More plants are now exploring ways to recover wastepaper, polymers that do not meet specifications, wood chips, bark, and similar wastes. The use of these wastes is made easier if the plant has a coal- or wood-fired boiler, but is complex if the original design did not provide for it. Collecting and handling the solids can have a significant cost.

In a number of places, centralized "refuse-to-power" plants are being constructed, which improve the economics, systematize the collection, and move the corrosion problems outside the plant gates. Obviously, individual plants must share the value of their wastes to take advantage of these benefits.

Problem 8-1

The diagram (page 161) shows the available heat contained in a flue gas resulting from the combustion of a refinery gaseous fuel. Such diagrams can be constructed for any fuel, and will differ with the composition and carbon-hydrogen ratio of the fuel. From the chart the heat available to the process from the combustion of an amount of fuel can be calculated for a range of flue-gas conditions.

1. Calculate the fuel required to provide 100 million Btu/hr to process if the flue gas is at 1,000°F and combustion takes place with 200 percent excess air.

2. How much energy could be saved if the stack temperature were reduced to 300°F?

3   What would the total energy saving be if excess air could be reduced to 20 percent in concert with reducing stack temperatures to 300°F?

4   What process conditions would limit the extent to which efficiency improvement could be realized?

Solutions

1. Arrow 1 on the accompanying diagram (page 162) indicates that the gross efficiency at the specified flue-gas conditions is 37 percent. Therefore,

$$\text{Fuel required} = \frac{100 \times 10^6}{0.37} = 270 \times 10^6 \text{ Btu/hr.}$$

2. If stack temperature is reduced to 300°F, then arrow 2 describes the available heat. The efficiency is increased to 77.5 percent. Fuel required is

$$\frac{100 \times 10^6}{0.775} = 129 \times 10^6 \text{ Btu/hr.}$$

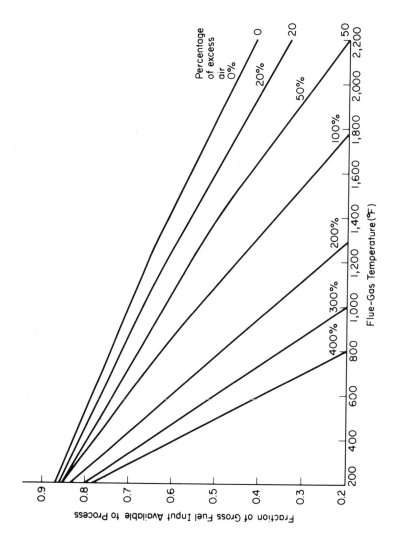

Diagram for Problem 8-1

Percentage
of excess
air
0%
0
20%
20
50%
50
100%
200%
300%
400%

Flue-Gas Temperature(°F)

Fraction of Gross Fuel Input Available to Process

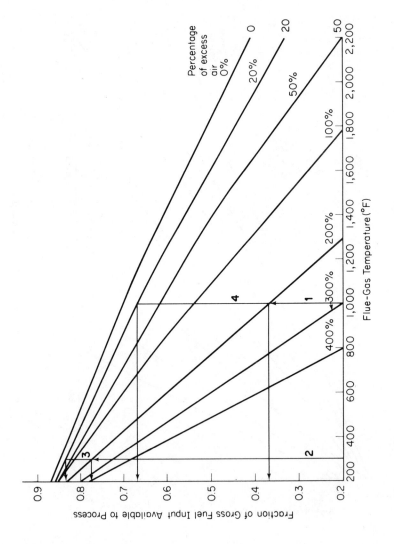

Diagram for solutions to problem 8-1.

Savings are

$270 - 129 = 141$ x 10^6 Btu/hr.

3.   Further efficiency improvement (arrow 3) is possible by reducing excess air. The efficiency of the combined improvement is 83.5 percent. Fuel required is

$$\frac{100 \times 10^6}{0.835} = 120 \times 10^6 \text{ Btu/hr.}$$

Combined saving is

$270 - 120 = 150$ x 10^6 Btu/hr.

4.   The efficiency of the process step can be doubled, mostly by reducing flue-gas temperature. The heat made available is between 300°F and 1,000°F. The key to implementing such a project from the process standpoint is to find a use for the lower heat. If it cannot be used directly in the process, air preheat for the combustion process and/or steam generation may be feasible. Alternatively, the stack temperature can be left alone and the excess air reduced. Arrow 4 shows that efficiency can be increased to 67 percent in this way. Fuel required is

$$\frac{100}{0.67} = 150 \times 10^6 \text{ Btu/hr.}$$

Savings are

$270 - 150 = 120$ x 10^6 Btu/hr.

TECHNIQUE FOR RECOVERING WASTE HEAT

Value of Recovered Energy

In an industrial environment the recovery of energy is driven by economics. Even if state or federal governments mandate energy efficiencies or cutbacks, economics will still be the driving force, since the alternative of ceasing to produce is always available.

The economic value of recovered energy is set by the market value of the resource it saves, such as fuel, electric power, or some other energy form. It follows that an understanding of where the recovered energy fits into the overall energy-supply system of the plant is essential for economic analyses.

The classic, and all too common, example of tunnel vision on the part of well-meaning energy conservers involves the unit supervisor or engineer who wants to improve his unit's efficiency. He devises a project to recover waste heat as low-pressure steam (usually the cheapest way) and reduces his unit's consumption or even exports steam to a nearby header. He is often aided in developing an economic project by a steam pricing system based on enthalpy, which provides him a high credit for the steam produced. After startup, a vent valve in the utility plant opens wider, and the profits from the investment go up the stack.

In general, the most profitable use of recovered energy is for low-temperature applications. Preheating combustion air or feed materials that are ultimately heated by fuel (including that supplied indirectly by the steam system at generation pressure) is the most direct approach. There are no interactions to be considered, and the results can be measured directly on the fuel meter (if there is one).

In plants that generate their own steam, the correct way to evaluate steam savings is to calculate the impact of the potential saving on the overall fuel used at the site, including the boilers. The true economic impact is thus calculated in easily measurable and consistent terms, that is, the fuel invoice. In the example cited, the saving would have been zero because of the steam system imbalance. There is no confusion over accounting steam valves based on costs; the net economic impact is clear.

The value of fuel savings may be enhanced by supply considerations. In many areas the supply of desirable gas and light liquid fuels typically burned in process heaters is limited and can affect production. Thus, fuel savings can be reflected in incremental production profits. The switch of demand from light process fuel to heavier, cheaper boiler fuel is often an additional economic incentive.

Power savings introduce another level of complexity in the evaluation problem. This is particularly true in plants that raise high-pressure steam and generate power in backpressure turbines exhausting to process-steam pressures and in condensing turbines. Often such plants also purchase backup electricity. In some parts of the country, purchased power is cheaper than incremental in-plant generation from condensing turbines. Purchased power is most often cheaper when the utility has long-term fuel contracts at prices below current plant fuel values, when appreciable capacity is supplied by hydroelectric stations, or when off-peak power can be utilized in the plant. Thus, while electricity is pure work and thermo-dynamically the most valuable utility, saving it can have various economic impacts. There is no substitute for a correct economic analysis.

IMPORTANT METHODS OF RECOVERY

Flue-Gas Heat Recovery

In some cases there is still room to use flue gas directly in the process concerned or in an adjacent one. This generally provides the opportunity for greater fuel savings than does preheating of combustion air. Consider the example illustrated in Figure 8-2. In this process, there is no heat of reaction. Of the 100 fuel units fired, 40 heat the feed to the desired temperature, 10 account for radiation and leakage losses, and the remaining 50 leave as sensible heat in the 1,600°F flue gas. This flue gas represents a major opportunity for energy recovery.

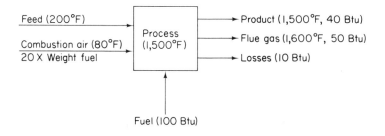

Figure 8-2. Base case.

Using some of the flue gas to preheat feed by adding a convection section to the furnace, we can reduce the required firing. As shown in Figure 8-3, recovering 12.5 units of energy from the flue gas reduces fuel requirements by 25 units. As the available-heat diagram in unit 1 shows, reducing the stack temperature to 1,000°F increases furnace efficiency to 67 percent. Supplying the required 50 Btu of energy at 67 percent efficiency requires 75 Btu of fuel to be fired.

If feed materials are already being preheated, then flue gases can be used to preheat the combustion air used in the heater. For a typical natural-gas fuel, about 20 lb of air are required per pound of fuel burned at about 20 percent excess air. All this air must be heated from ambient to flame temperature, and then it exits as flue gas, which in turn can be used to preheat the air. A schematic of an air preheater is shown in Figure 8-4.

Figures 8-3 and 8-4 show that feed preheat and air preheat save the same amount of energy. Factors that differ are the capital costs of the two routes and the extent of recovery possible. The temperature of the preheated air will match that of the flue gas sooner than will the feed (greater heat capacity), so the feed preheat holds greater potential for energy recovery.

Obviously, other options to recover energy from the flue gas can be envisioned, including direct integration with other processes and steam generation.

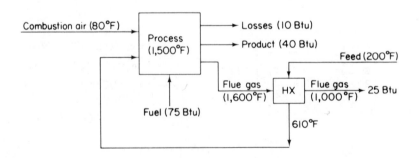

Figure 8-3. Preheat feed.

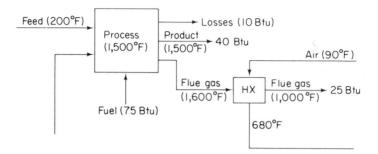

Figure 8-4. Air preheat.

In dryers, paper machines, and other operations using very large air flows to transport heat, process constraints may set the extent of possible energy savings. The operation of a paper machine depends very much on the humidity of the incoming hot air. Obviously, direct recycle of warm wet exhaust air is impossible, but indirect exchange with cool dry air could reduce the steam requirement. See Figure 8-5.

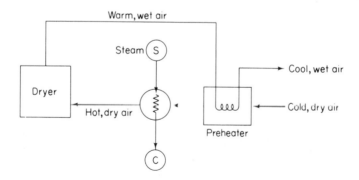

Figure 8-5.

Gas turbines also use very large excess-air rates (200–300 percent) and high (800°F) exhaust temperatures for process reasons. The mass of hot air is a key factor in generating power in the turbine. However, all this air must be compressed to the pressure of the combustion chamber, and compression is most effectively carried out at low temperature (lower volume). Direct exchange of hot exhaust air against cool inlet air thus has a detrimental effect on the net horsepower produced by the machine. Since gas-turbine exhaust contains about 17 percent oxygen, it is often used as hot combustion air for boilers or furnaces. More frequently, fired or unfired waste-heat boilers with low pressure drop are placed in the exhaust duct of gas turbines, and their net production is integrated with the overall steam power system. General Electric and other manufacturers provide surveys showing steam-generation possibilities by means of this method.

Direct Heat Integration

In many cases, energy is rejected in condensers or coolers because at the time of the original designs it was not economic to provide the extra heat-exchange surface needed to recover it. At present and projected future fuel costs, it is often economical to match process energy needs with heat being rejected.

A brief example of multiple-effect heat use in a petrochemical process follows. In the recovery of pure aromatic chemicals from refinery reformate streams, distillation is generally used to separate xylenes from benzene and toluene. In many cases, separate reboilers are provided for each tower, as indicated in Figure 8-6a. With small or negligible adjustments in the pressure of the xylene splitter, the overhead can often be used to reboil the benzene tower, as shown in Figure 8-6b. Adequate duty is maintained on the original condenser to ensure stable pressure control on the xylene splitter. At the required operating temperatures, it is appropriate to consider using the benzene tower condenser or the xylene tower bottoms cooler to provide feed preheat for these or other towers in the complex.

Steam Generation

The site steam balance will control the economics of using waste heat to generate steam. Equipment is conventional and often arranged in convenient packages by vendors. Generally, the higher the pressure of steam generated, the better. Despite this, one pays for the flexibility inherent in recovered steam in the capital cost for the system, and in an energy value (generally lower than fuel), unless

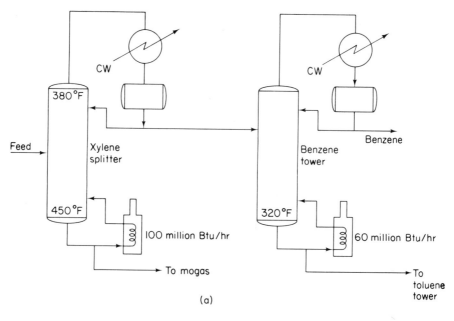

(a)

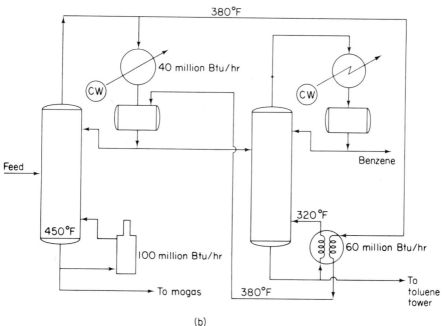

(b)

Figure 8-6.

fired boilers are directly reduced in throughput. In addition, an appropriate supply of boiler feedwater must be provided if it is not already available at the recovery site.

Gas turbines and furnaces with comparable stack temperatures are primary candidates for waste-heat boilers. Occasionally it is more practical to recover steam than to go to extremes to affect direct integration of process heat, because of steam's ease of transport, widespread use in the plant, and well-defined operating practices.

Space Conditioning

Plants with large needs for space conditioning (or neighboring plants with the same needs) may have some opportunities for improved energy efficiency. Either heating or cooling may be economical if sources of waste heat are located close to needs. Low-pressure steam is costly to transport, and no one will bring potentially messy or hazardous process streams into high-occupancy areas, but using it to heat warehouses, finishing buildings, tankage, and even neighboring greenhouses has been shown to be economical.

In summer, when low-pressure steam may be vented, absorption refrigeration may be economical. Many buildings use lithiumbromide systems for space cooling. A schematic flow diagram for a typical system is shown in Figure 8-7. About 1 Btu of chilled water can be produced from 1.5 Btu of waste heat at around 220°F. More flexibility can be obtained with ammonia absorption units at about 20 percent higher capital costs. In either case a cooling medium must be available for all of the waste heat.

The economics of absorption refrigeration are very much subject to local circumstances. Truly low-cost heat is required, and small size (< 20 million Btu/hr) refrigeration is often an incentive compared with mechanical refrigeration.

The export of either heat or refrigeration is subject to many institutional problems. Many community-oriented projects function well in places such as Iceland and Sweden, but the United States has shown little inclination to foster similar efforts at combined energy efficiency. If the federal government provides incentives for cogeneration, a first step in the potentially immense market may be possible.

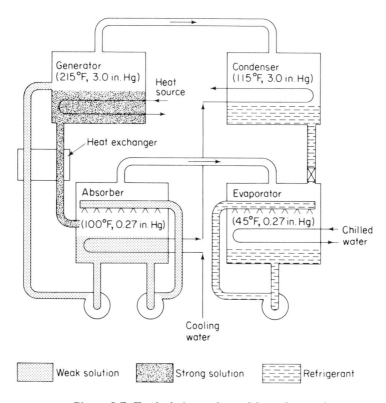

Figure 8-7. Typical absorption refrigeration cycle.

Direct Recovery of Work from Letdown
of High Pressure Fluids

In general, machinery is lacking for true multiphase applications, but all-liquid and all-vapor systems have been used. Steam is the most common vapor used, and generally offers opportunities for energy savings. As power costs rise, letdown valves are being replaced by backpressure turbines attached to process or utility loads. In new applications, small (200–500 hp) multi-stage turbines, which have an efficiency of 50–60 percent, can be justified versus the usual single-stage unit with an efficiency of 25–30 percent. This is often true not because of the higher power generated, but because less steam is required for a fixed power demand, and steam-balance improvements result.

Power Generation Using an Organic Rankine Cycle

The Rankine cycle is the same as the one used in steam power plants. Here, however, the working fluid is a special organic compound in place of water. Such fluids can improve cycle efficiencies at low (250–550°F) heat-source temperatures, and permit economical power recovery.

A typical thermodynamic cycle is shown in Figure 8-8. Pressures and temperatures will vary with the working fluid. Several vendors offer packaged systems using Freon, toluene, and the like. Some efficiencies for a hexane system versus water are shown in Figure 8-9.

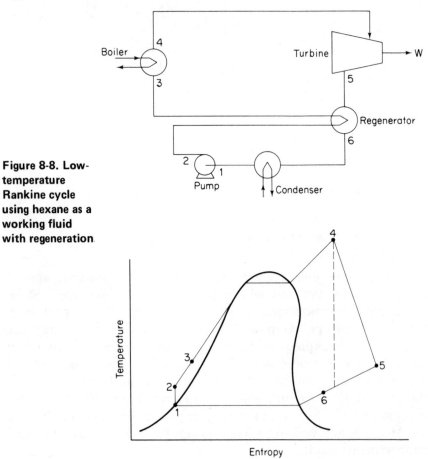

Figure 8-8. Low-temperature Rankine cycle using hexane as a working fluid with regeneration

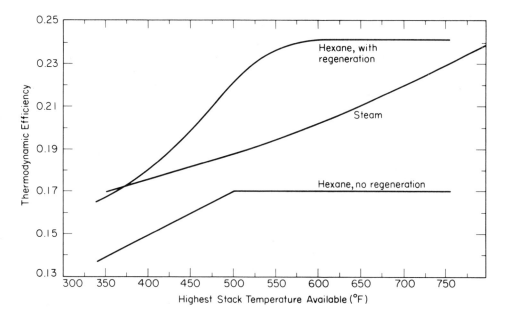

Figure 8-9. Work-recovery potential with condensing at 110°F, 75% turbine efficiencies, and stack gas cooled to 300°F.

GENERAL CONSIDERATIONS THAT NARROW CHOICES

Steam Balance

There is no profit in generating steam at one unit and vending it at another. Steam savings must be reflected in fuel savings somewhere in the plant to be valid. A sure approach if the steam system is in reasonable balance is to couple a new steam use with a generation project. The net effect on the overall steam balance is thus zero.

In many cases, low-pressure steam is in excess at least part of the year. This can hurt the economics of many steam-recovery projects and may indicate that other means of recovery should be given higher priority. An overall plan to rebalance the steam system should be developed before decisions are made on specific projects.

Layout

The capital cost of large hot-air or flue-gas ducts or low-pressure steam lines limits the economic length of recovery paths. The same is also true of high-pressure superheated steam lines. Interunit transfer of energy in this type of medium is rarely economical. Flue-gas air preheaters are rarely successful unless the hot air can be used in the same block. Low-pressure steam can rarely be transported off site profitably unless volumes approach ½ million lb/hr. Steam at greater than 900°F requires alloy piping and is generally used to drive turbines. Distances of more than 1,000 feet incur rapidly escalating capital costs. On the other hand, steam at 600 psi and 750°F can often be transported 1–5 miles to replace fuel.

Safety

In mature units, congestion is often a serious problem. Large pipes or heat exchangers may restrict access for fire fighting, maintenance, or emergency egress routes. Large liquid holdups may greatly exacerbate potential fire hazards, especially if space for isolation valves is not available. Encroachment on personnel access ways may lead to burns and bumps, causing disabling injuries or significantly reduced operator performance. Thorough review of proposed layout changes by safety committees and operations groups is essential. In some cases scale models may be needed to present the situation clearly and to gain the commitment of an operations group to the project.

Operability and Service

Enhanced energy efficiency is often paid for, in part, by reduced operating flexibility. Consider the xylene splitter and benzene tower discussed previously. In the original configuration, heat input to each of the two towers was independent and expandable at the price of furnace efficiency by overfiring the furnaces. Once the two towers have been coupled by using the overhead of the xylene splitter to reboil the benzene tower, heat inputs are no longer independent. In addition, pressure control on the first tower affects heat input to the second. Temperature driving forces on the second tower are reduced, and they are restricted by process conditions

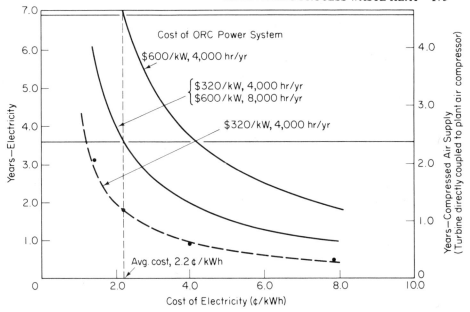

Figure 8-10. Payback period for organic Rankine cycle power system.

on the first tower, not by the capabilities of the fuel system. In short, the operators are working within narrower limits to maintain product specifications.

Interunit integration compounds operability questions. In a number of cases direct integration has been practiced across unit boundaries. This makes the energy source for one unit subject to the service factor on another unit. In general, revamp situations maintain independence by preserving the original energy supply and rejection arrangements, which can be reverted to in an emergency. Safety considerations or layouts sometimes prevent such insurance, and reductions in annual capacity must be considered in integration economics. "Grassroots" installations face similar questions.

Heat pumps or other machinery-aided heat recovery systems add a degree of complexity to the unit. Preserving the original energy supply path is frequently impossible in these systems, and lower service factors must be considered.

Seasonal effects may enhance or detract from project economics. If a light ends tower is in summer by condenser capacity, a heat pump may eliminate the problem. It may even add to capacity by lowering the total pressure, thus lowering reflux requirements. On the other hand, increased heat losses in winter may test compressor capacity, since temperature driving forces are small. More careful analysis is important in any case.

Problem 8-2

Consider the furnace shown in the diagram (page 177). Fired duty is 100 million Btu/hr, and stack oxygen is 8 percent but radiant-section oxygen is 3 percent. List potential steps to improve furnace efficiency in two categories: (1) operating improvements (negligible investment) and (2) facilities improvement. Calculate potential savings. Comment on economic priorities for all steps. (See appendix B for con-conversion of flue-gas oxygen content to the percentage of excess air.)

Solutions
Negligible Investment Possibilities

With 3 percent oxygen at the radiant-section exit, burners are well tuned. Leakage across the convection section is very large and could be rectified by simple techniques. Aluminum tape and/or fibrous mastics can be applied. If necessary, cover plates and/or gaskets can be replaced. No more than 1 percent change in oxygen content is generally experienced in well-maintained furnaces.

The savings in fuel from reducing excess oxygen from 8 percent to 4 percent can be approximated from the available heat diagram (see Problem 8-1). For natural gas, 8 percent oxygen represents 55 percent excess air and 4 percent oxygen corresponds to 21 percent excess air. The efficiency at 600°F and 55 percent air is 73 percent, so process duty in the furnace is 73 million Btu/hr. Reducing excess air to 21 percent increases efficiency to 76.5 percent. Thus, the fuel requirement is $73/0.765 = 95.4$ million Btu/hr and the savings are about 4.5 million Btu/hr.

Investment Options

If excess oxygen is reduced as above, stack temperatures can be reduced to save additional energy. Since the fuel is sulfur-free natural gas, a stack temperature as low as 250°F is possible. From the same graph, at 250°F, efficiency should be about 85 percent. The fuel requirement is $73/0.85 = 85.9$ million Btu/hr. The savings potential is thus an additional 10 million Btu/hr. Given only the information contained in the problem, either air preheat or steam generation would be appropriate to study. Feed temperature is too high to extract more heat in the convection section. In general, air preheat would be the first priority because it directly reduces fuel. Before developing air preheat in detail the designer should develop information on other possible uses for the stack heat.

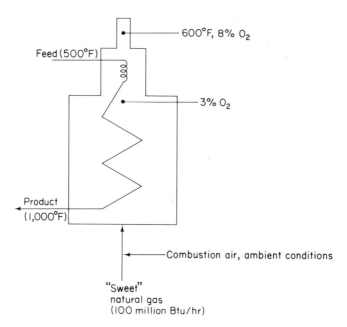

Diagram for Problem 8-2

MATCHING SOURCES AND SINKS

Current Economic Standards

When most existing plants were designed, fuel costs were substantially lower. Good designers optimized capital cost versus potential energy savings and arrived at configurations that are inefficient by today's standards. Not all of this difference can be recouped, because of inflation and the inherently higher cost of adding features originally omitted. However, a good understanding of current design standards can point the way to improvements in existing systems.

Table 8-2 lists some "grassroots" energy standards that are generally economical today. These can be applied to either a proposed new design or an existing plant to quantify inherent energy losses. By combining these general standards with specific process-related standards, an existing plant can be examined for potential facilities improvements. For example, if we say that the target steam-hydrocarbon ratio for an ethylene plant is 0.3 and that motor drivers are most efficient for the site, we can analyze in gross fashion the energy losses inherent in a proposed or existing steam cracker that deviates from these standards.

Table 8-2. Some standards for energy-efficient plants.

Stack temperatures of 250°F (121°C) or as limited by the economics of fuel sulfur up to the maximum of 350°F (177°C)

Excess oxygen in gas-fired furnace stacks of 2.0 percent and in liquid fuel-fired furnaces 3.5 percent (dry basis)

No steam leaks or venting to maintain steam balance

90 percent condensate return

No recycle around pumps and compressors

1,500 psig minimum steam generation pressure

No heat rejected to cooling water or air at temperatures greater than 250°F (121°C)

Minimum control valve ΔP on process streams

No condensing steam-turbine drivers

Minimum steam pressure for process heating

No pressure letdown across control valves

Fractionation operated at 1.1 times minimum reflux or as dictated by current economics of trays versus reflux

Economic insulation thicknesses based on current fuel and capital costs

No cooling and reheat of intra- and interunit surge capacity

Table 8-3 compares the design parameters of a 1960-vintage ethylene plant based on ethane feed against the suggested standards. Three areas of major potential are immediately obvious without any detailed process knowledge: furnace stack temperature, quench inlet temperature, and process-gas compressor driver. Such ideas

Table 8-3. Examples of energy loss for gas cracker (ethane feed; 105,000 met/yr C$_2$-C$_4$ olefins produced).

	1978 Standards	1960 Plant	Annual Losses versus Standard	
			Millions of Btu	Thousands of dollars @ $2 per million Btu
Cracking furnace stack temperature	250°F	700°F	427	299
Excess air	10%	15%	14	9.8
Steam-HC ratio	0.3	0.3	—	—
Temperature to quench tower	250°F	3 @ 600°F 2 @ 400°F	260.4	182.3
Process gas compressor driver	Electric	Condensing turbine		
Equivalent compressor fuel (millions of Btu per day)	1,174.1	1,904.6	730.0	511.0
Refrigeration drivers: C$_3$	Electric	Extraction condensing turbine		
C$_2$	Electric	Backpressure turbine		
Equivalent refrigeration fuel (millions of Btu per day)	1,287.8	1,328.2	40.4	28.3
Total (millions of Btu per day)	5,468.6	6,940.4	1,471.8	1,030.4
Percentage of 1978 Standards	—	126.9		

represent an initial phase of energy-efficiency improvement and can be evaluated relatively independently.

Of course, choices must still be made about where to use recovered heat most economically. The heat currently wasted in the furnace stack could be recovered as air preheat, or as steam, or perhaps as additional boiler feedwater or feed preheat. Not enough information is given here to make the choice in optimum fashion, but the major possibilities can be quickly evaluated.

A more complete analysis can be carried out with only a little more effort. Consider the flow plan in Figure 8-11. With the data presented, energy use can be characterized and possible matchups suggested. To do this, the amounts and required process conditions for energy inputs are compared with similar characteristics for rejected heat.

Table 8-4 lists the overall energy balance. The flow sheet does not supply all the data, but approximations are readily available.

Table 8-4. Overall energy balance.

Stream Location	Millions of Btu per Hour
Outputs	
E-1	3.5
E-2	5.5
E-4	0.04
E-5 [1,416 × 500 × (120 − 70) × 1]	39.4
Sensible heat of products versus feed	1.1
Subtotal	49.54
Furnace stack (by difference)	20.76
Total	70.3
Inputs	
Furnace fuel	58.0
E-3	13.3
Total	70.3

The following data were obtained from Perry's *Chemical Engineers Handbook*. With these data, the overall balance shown in Table 8-4 can be calculated. The furnace stack losses derived by difference can be checked with the available heat diagram in unit 1 (see test).

	Specific Heat (Btu/lb-°F)			
	50°F	150°F	185°F	Density (lb/gal)
Benzene	0.34	0.48	—	7.3
Toluene	0.364	—	0.534	7.2

Note: 1 bbl = 42 gal; 1 gal/min water = 500 lb/hr.

With a stack oxygen content of 10 percent, excess air is 81 percent. At a stack temperature of 830°F overall efficiency is 64 percent. The process duty is therefore 57 x 0.64 = 36.5 million Btu/hr, or the loss is 20.5 million Btu/hr. This balance is close enough to establish a list of possible waste-heat-recovery ideas.

Systematic Approach to Idea Generation

By combining this heat balance with the stated process conditions we are ready to generate a list of energy efficiency improvements. Table 8-5 lists heat inputs and rejections along with the required process conditions. A quick check suggests that the bottoms cooler, E-2, and/or part of the condenser, E-4, might be used to preheat feed to save at least 5 million Btu/hr. Product coolers E-1, E-4, and E-5 are small and two are at low temperature. These are probably not attractive opportunities in this unit, unless some of this energy might be recovered to preheat boiler feedwater if a need exists.

The next consideration is the reboiler furnace, F-1. Potential improvements fall into two categories: operating and improved facilities. Referring again to the available-heat diagram in unit 1, we find that furnace efficiency can be improved to about 71 percent by reducing excess air to 20 percent from the present 81 percent. This means fuel fired would be reduced to 36.5/0.71 = 51.4 million Btu/hr, for a savings of 5.6 million Btu/hr simply by furnace adjustment.

In the longer term, there are four ways to improve furnace efficiency. If we refer to the available-heat diagram again, we see that furnace efficiency could be improved to about 83 percent if stack temperatures could be lowered to 300°F. From Table 8-5 we see that feed preheat could be carried out with an additional convec-

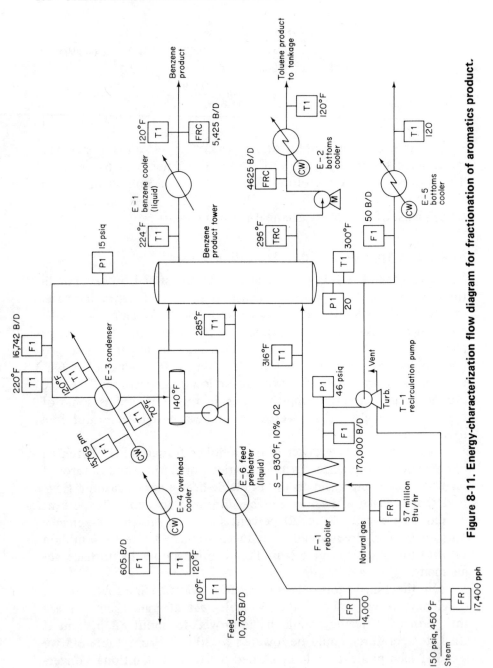

Figure 8-11. Energy-characterization flow diagram for fractionation of aromatics product.

Table 8-5. Comparison of process requirements.

	Duty (Millions of Btu per Hour)	Process Temperatures (°F)		Medium	Heat-Transfer Temperatures (°F)	
		In	Out		In	Out
Inputs						
F-1, reboiler	36.5	300	316	Fuel		830
E-6, preheater	13.3	100	285	Steam	366	366
Rejection						
E-1, benzene cooler	3.5	224	120	Cooling water	70	?
E-2, toluene cooler	5.5	295	120	Cooling water	70	?
E-3, condenser	39.4	220	140	Cooling water	70	120
E-4, overhead cooler	0.07	140	120	Cooling water	70	?
E-5, bottoms cooler	0.07	300	120	Cooling water	70	?

tion section. Of course, both of the "standard" approaches to furnace efficiency improvement—combustion-air preheat and steam generation—can be considered. In a more costly vein, the waste heat could be recovered to generate electricity. Finally, we note that the furnace may be eliminated entirely and the heat supplied with 150-psig steam.

Before completing our list, we should consider the turbine-driven recirculation pump. The theoretical fluid horsepower is calculated from flow and head as follows:

$$\frac{170,000 \times 42}{(24 \times 60)} \times 7 \text{ lb/gal} \times \frac{(35 - 20) \text{ lb/in.}^2 \times 144 \text{ in.}^2/\text{ft}^2}{33,000 \text{ ft-lb/(min/hp)} \times 52.6 \text{ lb/ft}^3} = 72 \text{ hp.}$$

From the conventional Mollier diagram [see *Steam Tables* (Windsor, Conn.: Combustion Engineering, Inc.)], we find that the isentropic expansion of steam at 150 psig, 450°F to atmospheric pressure should yield 192 Btu (0.0755 hp) of work per pound. The theoretical steam rate should be 954 lb/hr instead of the measured 3,400 lb/hr. This corresponds to an overall efficiency of 28 percent. A motor or a more efficient turbine should be considered.

The pump is probably running to protect the furnace from overheating during a power failure. (Someone feared the turbine would not start automatically when needed, or slow-rolled it so fast that it was decided to run the turbine as the normal driver.) If the furnace were eliminated, the pump could be either eliminated or returned to motor drive.

The turbine could also be discharged into a condenser pre-heating the feed at the cold end. This would lower the turbine back-pressure, recovering more work per pound of steam. Also, turbine efficiency would be less of a factor, since all the exhaust heat would be used, albeit at a low level. The investment for a condenser on such a small turbine is probably not justified. The idea list is recapitulated in Table 8-6.

ECONOMIC GUIDELINES

In many cases, equal savings can be obtained by different methods. Obviously, the least costly would be the best. Qualitative comparisons of capital requirements can often identify the most

Table 8-6. Summary of potential waste-heat recovery ideas.

	Potential Savings (Millions of Btu per Hour)
Operational	
Cut excess air to furnace	5.6
Investment	
1. Improve furnace efficiency	
a. Preheat feed in convection section	13
b. Raise steam	15+
c. Preheat air	8[a]
d. Substitute steam heater	(Δ efficiency $\times$ 38)
e. Generate power via Rankine cycle	3–4 (as power)
2. Recover heat rejected to cooling water	
a. Reuse toluene cooler heat to preheat feed	5
b. Reuse part of condenser heat to preheat feed	5–7
3. Save steam at turbine	
a. Replace with motor	
b. Improve efficiency	

a. From Figure 5.1.

attractive method without extensive cost-estimation efforts.

Consider the list in Table 8-6 from this point of view. Two sizes of heat-recovery projects are described: a small one for about 5 million Btu/hr and furnace heat recovery for 10–15 million Btu/hr. Of the smaller possibilities, use of the toluene cooler probably requires the least capital. At the cost of some piping, feed can probably be substituted for the cooling water and most of the heat can be recovered directly. Of course, the toluene will exit at a somewhat higher temperature than the present 120°F, unless the existing exchanger is oversized. Unless piping losses make up for lack of exchanger capacity, a small trim cooler may be needed. Use of the condenser heat will probably require a greater investment, because the existing exchanger must be retained in cooling-water service to handle the bulk of the duty. As a result, the entire preheat duty must be provided in a new exchanger. Thus, project 2a should be compared first with the larger projects.

For furnace-efficiency improvement we note that projects 1a, 1b, and 1c recover about the same amount of heat. Very roughly, the facilities needed to carry out each of these projects can be compared to shed some light on relative capital costs. See Table 8-7.

On this qualitative basis, direct feed preheat appears to be the best possibility, because capital costs will be low. Slightly more heat recovery may be possible with the waste-heat boiler, because not all the available heat in the flue gas is needed to preheat feed; but incremental capital costs are required. Alternatively, the air-preheat project also has the benefit of directly reducing fuel, while the value of steam generated must be determined from the plant steam balance.

For electricity generation, capital costs are generally in the range of $1,000 per kW. Qualitatively, this system has higher cost for heat exchanger, machinery, and structure than any of the other cases, and comparable costs in other areas. The attractiveness of this investment depends on the relative prices of electricity and fuel.

The plant program thus boils down to deciding whether both feed preheat and furnace efficiency can be justified. The most likely cases are either 1a, or 2a and 1c, unless further process, layout, or safety considerations decree otherwise. These matters must be checked before large-scale commitments to engineering and manpower are entered into.

Table 8-7. Facilities required by projects 1a-1d of Table 8-6.

	Preheat Feed (1a)	Make 150-psig Steam (1b)	Preheat Air (1c)	Generate Power (1d)
Heat exchanger	Base	≥ Base	> Base	≫ Base
Flue gas ducts	?	Large	Large	?
Forced-air fan	No	No	Yes	No
Induced-draft fan	Yes	Yes	Yes	Yes
Boiler feedwater supply	No	Yes	No	No
Vapor drum	No	Yes	No	Yes
Foundations or structure	Base	> Base	> Base	≫ Base
Electrical supply	Base	Base	> Base	≫ Base
Machinery	No	No	Minor	Major

The furnace-replacement project deserves special consideration. The capital cost is likely to be high, because of the need for a large heat-exchange surface. Savings are limited to the difference between the boiler efficiency (85–90 percent) and the 64 percent furnace efficiency (at most, 9–10 million Btu/hr). However, such a change may provide the opportunity to switch all of the duty from natural gas to liquid or solid boiler fuel. In addition, a thermosyphon would eliminate a pumping service. In these days of shortage such a switch may have site-specific product or capital credits.

To summarize likely priorities, plant programs generally fall into the following order of descending economic incentive:

- direct reuse of explicitly rejected heat,
- recovery to save fuel directly,
- recovery to generate steam,
- machinery-aided recovery.

Natural-gas phaseout or other fuel-switch credits may alter these relationships at any specific site. Recovery of quantities less than 1 million Btu/hr is not usually economical.

Specific consideration may also be necessary for the turbine-replacement possibilities. Venting turbines are generally relegated to spares. Steam consumption could be cut to about 2,100 lb/hr at the cost of a single condensing turbine installation. For a relatively small additional cost, steam use could be cut to about 1,500 lb/hr by buying one of the small multi-stage turbines now being made available. In many cases these systems are justified if motor drives are barred for safety reasons, but the small size of this driver probably precludes an economical substitution. In the horsepower range of about 200, practical projects may be possible.

Problem 8-3

Suggest all the possible ways to improve efficiency when integrating the process furnace with the steam power system shown in the diagram (page 189). (The furnace is as described in Problem 8-2.) Assign priorities to the ideas generated.

Solution

Losses occur in the process furnace stack and in the dump and boiler auxillary condensers. (The boiler itself is relatively efficient.) The magnitudes of these losses are as follows:

- — furnace stack over 300°F: 8 million Btu/hr (see available-heat diagram in the test for unit 1),

- — dump condenser: 0.10 x 300,000 x 857 Btu/lb = 25.7 million Btu/hr.

Possible uses for this energy include air preheat for furnace, air preheat for boiler, and waste-heat boiler on process furnace.

The optimal solution is complicated by the distance between the process furnace and the utility system.

Of themselves, steam-balance considerations militate against the waste-heat boiler concept. Additional steam generation will merely increase the load on the dump condenser. Air preheat becomes the logical application.

Geography also plays a role in the possibilities. Transporting 600°F flue gas for a mile to preheat boiler combustion air is impractical. These large ducts must be kept extremely compact, or the economics of recovery quickly disappear. Direct air preheat is thus limited to the process furnace alone.

The alternative is to use the steam system to transport the recovered energy. Steam at 150 psig has a saturation temperature of 366°F. The lowest stack temperature possible would be 400°F, and the savings would be reduced to two-thirds of the 8 million Btu/hr quoted above. Thus, we have two possible air-preheat projects: the process furnace alone and both furnaces using the steam system.

Combustion-air requirements for the furnaces can be approximated as follows. (Assume 20,000-Btu/lb fuel.)

- Fuel requirements are 100,000,000/20,000 = 5,000 lb/hr for the furnace and about 15,000 lb/hr for the boiler.

- Air requirements are about 20 lb per pound of fuel.

- Furnace air requirements are 100,000 lb/hr.

- Boiler air requirements are 300,000 lb/hr.

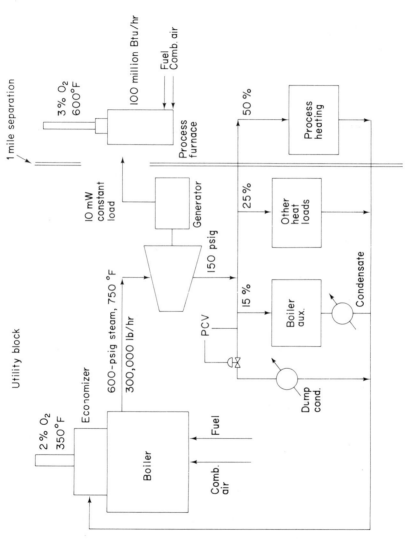

Diagram for Problem 8-3.

Preheat duties as a function of temperature are assuming 70°F ambient air) as follows.

- 100,000 x 0.24 x $(T_{final} - 70)$ = 24,000$(T - 70)$ for the furnace and 72,000 $(T_{final} - 70)$ for the boiler at 300°F combustion air temperature.
- Q_{FA} = 24,000(230) = 5.5 x 10^6 Btu/hr.
- Q_{BA} = 72,000(230) = 16.5 x 10^6 Btu/hr.
- Total = 22.0 x 10^6 Btu/hr.

(Somewhat higher air temperatures are possible for the direct furnace case.)

The question now becomes whether it is more economical to recover stack heat or use the dump condenser to provide all the combustion-air preheat. At the moment, no additional savings in energy will be possible by recovering stack heat. Qualitatively, we can compare the capital costs as follows:

	No Stack Recovery	Stack Recovery
Air preheat exchangers	Base	About same
Flue-gas duct	No	Yes
Induced-draft fan	No	Yes
Structures	Base	> Base
Instruments	Base	> Base
Piping	Base	≤ Base

Without more detailed examination of header sizes, pressure drops, equipment layouts, and so one, the logical project to consider is the use of the heat currently rejected in the dump condenser to preheat the air for both furnaces, because this can be done at lower capital cost than recovering stack losses from the process furnace. Future addition of a waste-heat boiler would still be possible.

PREHEATING COMBUSTION AIR OR FEED MATERIAL ESTIMATING SAVINGS

The amount of combustion air has a marked effect on combustion efficiency. In most combustion operations, preheating the air and fuel with flue gas or waste heat has significant potential for improved efficiency. This applies to the typical furnace with its low rate of excess air as well as to the dryer or solvent-removal unit which uses very large amounts of air (or inert gas) to evaporate (or calcine) a volatile material from a solid product.

Figure 8-12 is a nomograph allowing estimation of the savings that can be achieved by preheating air with flue gases from an industrial furnace. The graph was calculated for a typical distillate fuel at 20 percent excess air in a furnace operated 8,000 hours per year. The example shown considers a furnace with a process-heat duty of 150 million Btu/hr and a flue-gas temperature of 850°F.

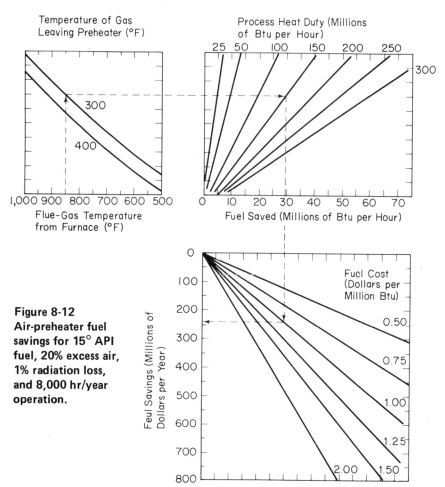

Figure 8-12
Air-preheater fuel savings for 15° API fuel, 20% excess air, 1% radiation loss, and 8,000 hr/year operation.

Entering the chart at the left at 850°F and assuming we would cool the flue gas to 300°F, we project horizontally to the 150 million Btu/hr line in the next section of the chart. Projecting to the abscissa, we find potential savings of 50 million Btu/hr. On the bottom chart in Figure 8-12 the annual value of these savings is calculated for various fuel costs. In the case shown, it is about $260,000 per year at $1 per million Btu; obviously, this saving escalates directly with fuel value.

The saving can be realized by means of either direct or indirect exchange. The choice depends on economics, but also on safety, available draft, plant layout, operating traditions, and other factors. In addition, preheated combustion air changes flame characteristics and flashback tendencies in premix burners, and increases the duty in the radiant section of the furnace. Using preheated air also increases the concentration of nitrous oxide in the flue gas by 50-100 ppm unless changes are made in combustion. Thus, mechanical and process factors need to be considered to establish the technical feasibility of air preheat before any further economic studies are made. The importance of these factors varies from application to application. In some applications, such as most boilers, these factors are unimportant; in others, such as reformers and steam crackers, they are very important.

TYPES OF PREHEATERS

Generally there are three practical systems for air (carrier gas) preheat:
- direct exchange with flue gas,
- direct exchange with another source of waste heat, and
- indirect exchange with flue gas.

In addition, feed materials can also be preheated with similar results, but this really represents adding a convection section to a furnace or a recuperator to a dryer. Some of the equipment and systems described in this unit are applicable to such tasks, and must be explored before the energy-conservation engineer considers more costly recovery schemes, such as combustion-air preheat.

Figure 8-13 shows a schematic of a typical direct-exchange system. The main components of the system are as follows:

- A forced-draft fan forces combustion air through the system to the burners.

- An air preheater exchanges heat between the flue gas and the combustion air. The two most common types are the regenerative and tubular.

- Air ducts transport combustion air from the forced-draft fan, through the air preheater, to the burners.

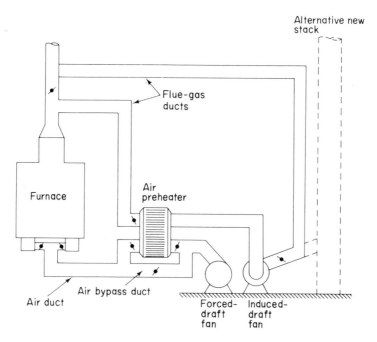

Figure 8-13
Schematic of typical air-preheater system.

Several types of heat exchangers may be used. The most common is the rotary (Ljungstrom) type for liquid or dirty gaseous fuels. The Ljungstrom preheater, illustrated in Figure 8-14, consists of metallic elements that are alternately heated and cooled. The metallic elements are contained in a subdivided cylinder that rotates inside a casing. Hot flue gas flows through one side of this cylinder and heats the elements, while the air to be heated flows through the other side. The cylinder rotates, and heat is transferred from the heated elements to the cooler air. The unit can be installed either vertically or horizontally.

Baffles that subdivide the cylinder, as well as seals between the cylinder and the casing, limit the amount of leakage from the air side to the flue-gas side. Since the air side is at a higher pressure than the flue-gas side, leakage is always toward the flue-gas side. This leakage, which is usually 10-20 percent of the total air flow, must be taken into account in the design of the preheater system, particularly the fans. Recommended margins are 25 percent on flow and 50 percent on head.

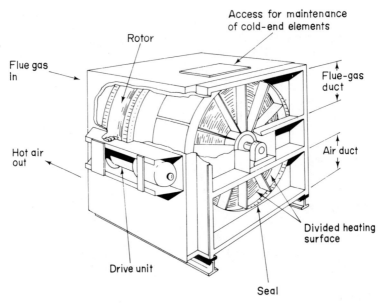

Figure 8-14
Ljungstrom-type regenerative air preheater.

The Ljungstrom regenerative air preheater, which is manufactured in the United States by Air Preheater Co., Inc., offers the following design features.

- A corrosion-resistant cold-end surface, which allows operation with outlet flue-gas temperatures of approximately 300°F. This surface, contained in baskets, can be easily replaced with a turnaround time of approximately one day. Cold-end material life has been in excess of 5 years. Note that the destruction of the cold-end elements by corrosion affects only the heat-transfer ability and not the operability of the preheater.

- A viewing port and light for onstream examination of the cold-end surface.

- An integral soot blower and connections for water washing facilities. For most fuels blowing is sufficient for cleaning the air preheater. Soot blowing is normally done once per shift with the preheater in operation. However, with extremely dirty fuels, automatic water-washing facilities may by added. Although water washing can be done with the preheater in service, it is best bypassed during this operation to avoid water carryover into the air ducts.

For clean fuels, with which soot formation and sulfur corrosion are not major problems, a tubular type of air preheater is an acceptable choice. It may also be less expensive than the regenerative type.

For retrofit applications the tubular air preheater consists of a tube bundle with extended surface (usually fins) located at grade. Normally, the flue gas flows across the tube bank while the combustion air flows inside the tubes. Since tubular air preheaters are individually engineered, they can be fabricated with almost any flow configuration desired. Figure 8-15 shows a typical flow arrangement, with the flue gas in crossflow and the combustion air making three passes through the tube bundle. At present, manufacturers of heat pipes (for example, Isothermics and Q-Dot) are offering combustion-air-preheat models which are more compact and less subject to leakage than conventional tubular exchangers.

An important consideration when specifying or purchasing tubular preheaters is cold-end corrosion. If the fuel has any sulfur content at all, alloy materials are required for corrosion protection with metal temperatures below approximately 300-350°F. On the basis of this corrosion consideration, flue-gas outlet temperatures are normally kept above 350-400°F. Should corrosion advance to the point where leakage occurs, the preheater has to be taken out of service because of the loss of combustion air by leakage. Depending on the construction of the preheater and the extent of corrosion, the correction may involve simply plugging the tubes, replacing tubes, or replacing the complete air preheater.

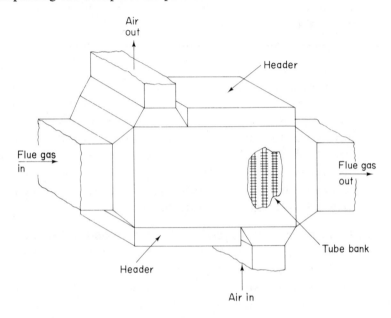

Figure 8-15
Tubular air preheater.

Another type of preheater that can be used is one transferring heat directly from a process fluid to the combustion air. The system shown in Figure 8-16 was recently installed on a chemical furnace. The heat source here is a recycle hot oil stream at 410°F. The preheater, an extended-surface tube bundle manufactured by Heat Research Co., was designed to reduce the stream temperature to 340°F while preheating the air to 375°F.

Because a hot-oil leak into the air plenum could produce a flammable mixture, special safety precautions are required. The safety features incoropoated in the design were an air-duct hydrocarbon leak detector, an alarm for low fuel flow, a design pressure of ten times operating pressure, the double rolling of tubes into the tube sheets, and the shutdown of oil flow to preheater if fuel is shut down.

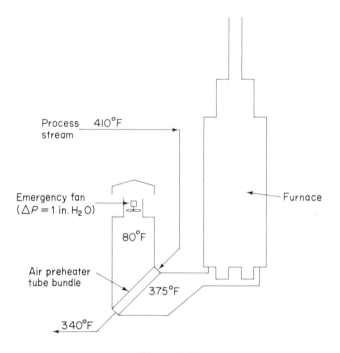

Figure 8-16
Direct-process air preheater.

    This type of air preheater requires much less space than a direct flue-gas-and-air system because the flue-gas ducts are eliminated. In many cases it can be operated on natural draft, but then it requires more heat-exchange surface.

    An adaptation of the design just cited can be used to effect indirect heat transfer from the stack to the combustion air. This type of unit, which has been patented by Heat Research Co., is shown schematically in Figure 8-17. Variations of this approach outside the patent have been used. The process stream becomes a heat-transfer medium. In this system adequate draft must be supplied to accommodate the increased convection-section pressure drop for the hot-oil tubes. All other things being equal, higher furnace efficiency is possible with this scheme because stack temperatures are reduced and the combustion air is preheated. In addition, multiple inputs and uses are possible without massive duct arrangements.

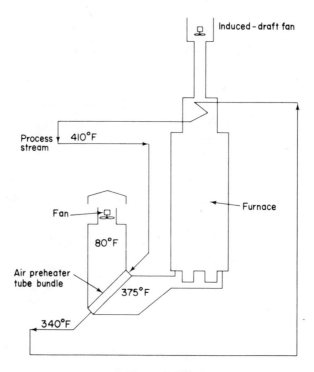

Figure 8-17
Indirect-process air preheater.

PREHEATING DRYER OR BUILDING AIR

In many operations large volumes of air are heated to dry either solid or film products and to provide conditioned ventilating air for finishing operations. Much of the hardware discussed previously is adaptable to these applications as well. Emphasis here is generally on tubular or heat-pipe types of exchangers, because fluids are usually clean.

In the simplest cases the heat can be reused either directly or indirectly. General Motors has implemented a number of reuse projects which have provided good returns on investment. In Figure 8-18 the use of flue gas from one furnace to supply the duty of a drying oven is described. In the case shown, flow requirements matched needs and the existing drying-oven stack provided enough draft. About 4 million Btu/hr were saved. Simple modifications to correct inadequacies in flow or draft are obvious. An induced-draft fan at point A could overcome pressure-drop problems. Also, this fan could induce flow of outside air at point B to control temperature and total flow rate.

The indirect use of flue gas is pictured in Figure 8-19. Here a compact heat-pipe coil is used to exchange about 4 million Btu/hr between flue gas and incoming air at low pressure drop. The indirect approach is indicated when flue-gas composition or contamination makes direct reuse impossible.

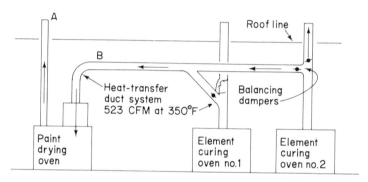

Figure 8-18
Direct use of flue gas.

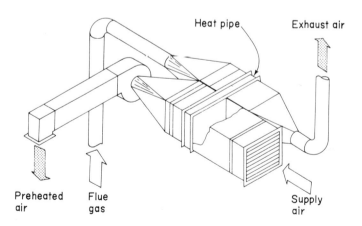

Figure 8-19
Indirect use for building or dryer heat.

Both of these items fall into the category of direct reuse, which should be the energy-conservation engineer's first priority. They are discussed here because the equipment involved is similar to air-pre-heat systems. An additional point is that even small projects (in the range of 4 million Btu/hr) can be economical if the right equipment and circumstances are combined.

In some cases further conditioning of the exhaust gas is neces-sary before process reuse is possible. This occurs in dryer applications where the humidity of the incoming warm air must be controlled to reach the desired product moisture content. In a paper machine, large quantities of warm dry air ($\sim$170°F, 2 percent relative humidity) are used to dry the slurry into paper. The exit conditions of the air are about 145°F and 60 percent relative humidity. Direct recycle of this air is impossible because of its moisture content. Extensive exchange with fresh incoming air (Figure 8-19) requires large amounts of sur-face and provision for full pressure drop for all of the incoming air.

In winter months the low absolute humidity of makeup air can make it possible to recycle some of the exhaust air and recover its heat at lower capital cost. Figure 8-20 describes winter conditions where half the air can be recycled. The amount of recycle air de-creases as ambient air temperature and humidity rise. Duty is thereby shifted from the heat-pipe exchanger to the steam coil to control the desired inlet-air conditions. In one application, where heat pipes were used to achieve a compact installation and to provide a surface on which water droplets could coalesce and be separated from the air stream, the payback period was about 2½ years.

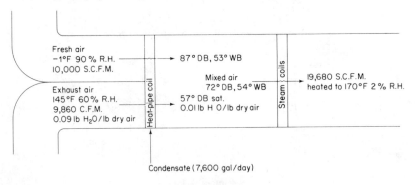

Figure 8-20
Conditioned recycle air to dryer; typical conditions at −1°F ambient.

EXAMPLES AND GUIDELINES

Consider the furnace shown in Figure 8-21. It is typical of furnaces used in major refineries and similar to the example discussed in Figure 8-12. Assume that preheating combustion air has already been determined to be the desired solution to improving furnace efficiency.

The types of air preheaters to consider are listed in Figure 8-22, in the form of a logic diagram. A simple process-fluid (or steam) air preheater was discarded at the outset, because none was available at the requisite temperature. Obviously, high-pressure steam could be used to provide comparable temperatures to a flue-gas or air preheater, but this would be a very expensive utility relative to waste heat from the stack.

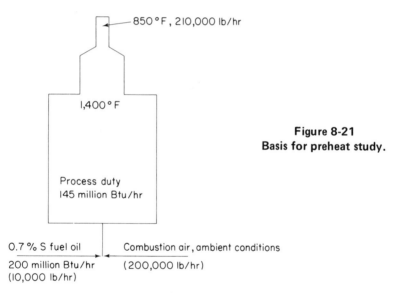

850 °F , 210,000 lb/hr

1,400 °F

Process duty
145 million Btu/hr

0.7 % S fuel oil
200 million Btu/hr
(10,000 lb/hr)

Combustion air, ambient conditions
(200,000 lb/hr)

Figure 8-21
Basis for preheat study.

The five remaining cases include two types of tubular preheater (Figure 8-15), a standard regenerative system (Figure 8-14), a heat pipe, and a self-contained intermediate fluid system similar to Figure 8-17. (Figure 8-23 illustrates the heat-pipe principle and a typical duct exchanger unit. Such packages are suitable for in-stack installation to eliminate much of the flue-gas ductwork needed with more standard units.) All possibilities were judged technically feasible to implement. The characteristics of the systems are compared in Table 8-8.

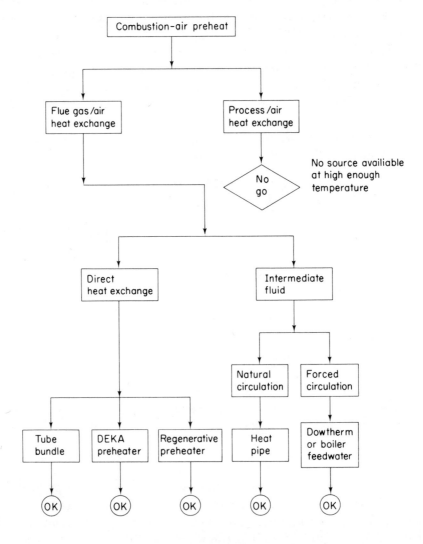

Figure 8-22
Logic flow diagram for scoping of air preheat alternatives.

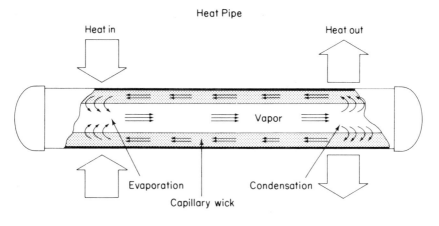

Heat Pipe

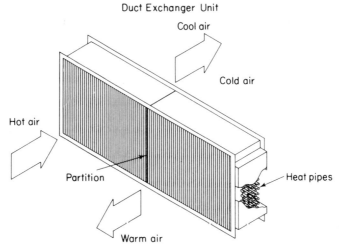

Duct Exchanger Unit

Figure 8-23

Two of the units are capable of withstanding more corrosive environments than the others. These are the regenerative type and the DEKA tubular unit. In the former, the rotating baskets can be coated and corrosion does not force shutdown. The DEKA unit is a massive collection of cast iron tubes bolted together. It has been proven in SO_3 environments. As a result, these units can recover slightly more heat than the others.

Other characteristics of the systems are compared qualitatively in the table. One of the key items involves the possibility that leakage of intermediate fluid could cause a fire, either in the stack or in the hot-air duct to the burners. The choice of working fluids and the desired safeguards are very important for controlling this potential hazard. The capital cost of the fluid unit includes the safeguards listed earlier for this type of preheater.

More than likely, a more qualitative consideration, such as service-factor history or experience with similar applications, will be important to the decision. In the author's experience, about half of all installations have ended up "regenerative" and half as "fluid" units. Heat pipes have been considered more seriously of late as experience has been gained. The tubular units have not been selected.

Table 8-8. Summary of probable air preheater economics.

	Fired Duty (Millions of Btu per Hour)		
	>60-80	30-40	<30
Direct regenerative air preheaters (Figure 8-14)	Yes	Marginal	No
Indirect fluid regenerative air preheaters (Figure 8-15)	Yes	Marginal	No
Waste process heat units (Figure 8-16)	Yes	Probably	Perhaps
Other special applications	Yes	Probably	Perhaps

Problem 8-4

1. List qualitatively the factors that would have to be evaluated in addition to economics in formulating a revamp air-preheater project.

2. List the advantages and disadvantages of intermediate-fluid types (Figure 8-17) versus the regenerative type (Figure 8-13).

Solutions

1. Depending on the local situation, the list of potential problems varies enormously. The following list represents the more general factors that must be considered:

layout, space

maintenance access

structures, foundations

stack-height limitations

service factor effects (need a bypass route)

variations in duties

seasonal effects

stack materials, corrosion

safety analysis, fire-fighting access

life of project

site fuel balance before and after

power supply for fans and preheater

local prejudice against induced-draft fans

fuel-composition variations

type of intermediate fluids available

turnaround schedules.

2. Relative to the regenerative type of preheater (Figure 8-13), the intermediate fluid type of preheater has the following advantages and disadvantages when the entire system in Figure 8-17 is required.

Advantages	*Disadvantages*
more compact	possible safety hazard for a leak
nonrotating heat exchangers	
positive seal	more surface in the two exchangers
no flue-gas ducts	
slightly lower cost	higher power consumption

STEAM GENERATION
ESTIMATING SAVINGS

Figure 8-24 is a nomograph similar to Figure 8-12 for estimating the savings possible by reducing the stack temperature of a typical furnace with a waste-heat boiler. It is based on a final stack temperature of 400°F to avoid sulfur corrosion, but it keys on heat fired in the furnace, not heat absorbed.

The arrows follow about the same case as Fig. 8-12: a furnace with an 850°F stack temperature firing 150 million Btu/hr. Reducing the stack temperature to 400°F can save about 16 million Btu/hr. The value of this heat depends on the steam conditions chosen and the temperature of the feedwater. The overall steam balance at the site must be evaluated to obtain realistic values for steam, unless the waste-heat boiler project is coupled with a new steam use.

To follow the example through, let us consider the low-pressure steam conditions of the pretest (150 psig saturated). The enthalpy of the steam is 1,196 Btu/lb. Boiler feedwater temperature is normally about 50°F below saturation to prevent localized boiling. This is 316°F. From the steam tables, the enthalpy of the water is then $341 - 50 = 291$ Btu/lb and the change in enthalpy, Δh, for steam generation is $1,196 - 291 = 905$ Btu/lb. Therefore 17,700 lb/hr of steam can be produced from the furnace if all the boiler-feedwater preheating is done outside the waste-heat boiler.

Corrosion from sulfur oxides in the fuel gas is a larger problem in waste-heat boiler applications than in air preheaters, because heat-transfer coefficients are higher for water than for air. This results in lower tube-metal temperatures for a given flue-gas temperature. In the case just discussed, preheating the boiler feedwater in the flue gas would probably not be possible without special considerations if the fuel oil contained any appreciable sulfur, because tube-metal temperatures would fall below the dew points of sulfur oxides. A rough guide for these dew points is given in Table 8-9. Common practice is to allow a 50°F margin for local variations.

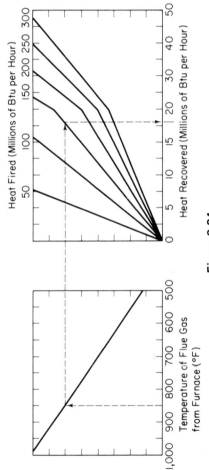

Figure 8-24.
Steam-generation heat recovery. Basis: 15° API fuel, 20% excess air, 1% radiation loss, 8,000 hr/year operation, 1,100 Btu/lb steam, flue gas at 400°F leaving steam coil.

Table 8-9

Relationship between sulfur content of fuel and dew point of sulfur oxides.

Sulfur in Fuel (Weight Percent)	Dew Point of Sulfur Oxides (°F)
0.2	220
0.5	260
0.7	275
1.5	300
> 2.0	340

TYPES OF WASTE—HEAT BOILERS

Three types of waste-heat boilers are in general use:

- specific designs for additions to furnace-convection sections or recovery of process heat,

- unfired (usually package) duct units, and

- auxiliary fired boilers for gas turbines and other systems that exhaust a high oxygen content (these may actually be conventional boilers using the gas turbine as a source of preheated air).

The first type includes a myriad of possibilities designed to capatalize on a specific opportunity, as well as relatively standard designs by vendors for a common process use, such as the waste-heat boilers on ethylene furnaces. Structural and draft limitations must be considered for most specific designs. General statements about this type of application are impossible, but most of what will be said about typical package systems will apply.

A schematic of a typical waste-heat-boiler installation is shown in Figure 8-25. The main components of this system are the waste-heat boiler, the induced-draft fan, and the flue-gas ducts. The induced-draft fan is required to compensate for the additional flue-gas pressure drop and the loss in natural draft due to the lower stack temperature.

The waste-heat boiler can be purchased from a wide variety of vendors of heat-exchange equipment or of gas-turbine waste-heat boilers, as well as from vendors of conventional furnaces. A typical vendor-designed unit is illustrated in Figure 8-22. The boiler consists of an extended-surface tube bank in an enlarged section of ductwork. Efficient arrangements for heating feedwater and superheating the steam are included in the package as appropriate. The vast majority of units are double-drum designs (Figure 8-26) and operate by natural circulation. The mud drum on the bottom serves as a tube header and settling area for solids in the system. The steam drum separates steam from water and provides the static head for water circulation.

Auxiliary-fired waste-heat boilers range from gas-fired duct types to large conventional boilers. Their purpose is to capitalize

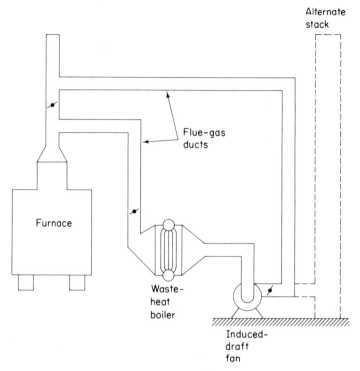

Figure 8-25
Schematic of typical waste-heat-boiler system.

on the presence of hot oxygen in the process stream. Enough additional fuel is fired to raise the temperature of the entire gas stream to the 1,400-1,800°F range. This both increases the heat recovered and raises the potential pressure of the steam produced. Most applications produce at least 600 psig steam, and many range up to 1,500 psig. Gas turbines are the principal heat source for auxiliary-fired units.

General Electric and other gas-turbine manufacturers have developed extensive data on the capabilities of waste heat boilers coupled to their specific models. Table 8-10 is an example of the kind of data available.

The vast range of possibilities is apparent. The flexibility of the auxiliary-firing option often allows easy adjustment to seasonal swings and boiler turnaround schedules. The combined cycle, where

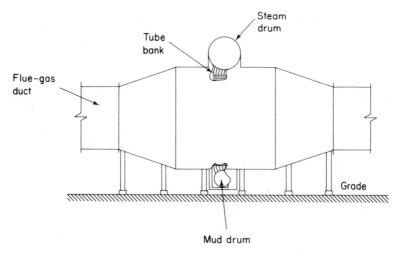

Figure 8-26
Typical waste-heat boiler.

the steam produced also drives a turbine, affords a very efficient way to generate power. According to General Electric, the fuel chargeable to power ranges from about 8,000 to 9,000 Btu/kWh in the unfired case to somewhat less in the fully fired case. A typical utility cycle needs 10,000 Btu/kWh. But remember that the utility is probably paying less for its fuel than an industrial complex.

Among the companies who have demonstrated experience in manufacturing waste-heat boilers in the United States are Struthers-Wells, Erie City, Vogt, Combustion Engineering, Babcock & Wilcox, Riley Stoker, Cleaver-Brooks, Born, and Enerex. Other companies also advertise capabilities in the engineering and construction of waste-heat boiler systems.

ECONOMIC CONSIDERATIONS

The economics of waste-heat boilers vary widely with steam values and with the extent to which existing facilities can be employed in the new system. In true "grassroots" installations the designer is generally free to arrange the steam balance, plant layout, and driver selection to make steam generation effective. As a result,

Table 8-10. Steam generation from exhaust of natural-gas-fueled gas turbine.

Turbine rating	10,400 kW
Fuel fired in turbine	147.7 million Btu/hr
Exhaust flow	388,800 lb/hr

	Unfired	Fired to 1,400°F	Fully Fired
Turbine output (kW)	9,290	9,200	9,050
Exhaust temperature (°F)	1,008	1,011	1,015
Auxiliary fuel (million Btu per hour)	0	48	333
	Possible Steam Production (lb/hr)		
150 psig, Saturated	67,100	—	—
420 psig, 655°F	55,200	96,400	304,000
630 psig, 755°F	51,400	92,100	296,000
895 psig, 830°F	48,500	89,100	290,000
1,315 psig, 905°F	—	86,500	284,000
1,525 psig, 955°F	—	84,300	

new gas turbines are almost always accompanied by waste-heat boilers, reformer furnaces (ethylene and ammonia) have their high-pressure waste-heat boiler on furnace effluents, and other processes use specialized process designs to conserve energy.

The energy-conservation engineer, however, must strive to improve an existing plant within the limitations imposed by the hardware already in place. Consequently, we shall address the retrofit problem in attempting to develop economic guildelines through some practical examples.

The first involves several old furnaces in a large chemical complex. These furnaces produced a combined flue-gas rate of 330,000 lb/hr at 1,000°F, currently vented through individual stacks. There is a local stack-height limit of 300 feet.

The flue gas rate corresponds roughly to a fuel rate 1,000 times higher—about 330 million Btu/hr. From Figure 8-24 we could expect to save about 50 million Btu/hr in a waste-heat boiler. Clearly, this project has significant potential.

Several cases must be studied in optimizing a scheme, but we will narrow these down to the two essential options: natural draft (Figure 8-27) and induced draft (Figure 8-28). Table 8-11 outlines the key variables for the cases considered. The natural-draft case is characterized by a lack of pressure drop. Heat recovery is limited by the need to maintain a stack temperature of 450°F and the lack of pressure drop for an economizer. The induced-draft case trades the capital and operating costs (and possible service-factor debits) of a fan for greater heat recovery and a shorter stack. Both cases use forced-circulation boiler feedwater rather than the natural-circulation model discussed earlier, and both produce 130-psig steam, mainly because of local preferences.

As illustrated in Table 8-12, generalizations about the economics of waste-heat-boiler projects are difficult to make. Convection-section additions are feasible at quite small sizes (10,000 lb/hr) if no major draft or structural additions are required. Major, multi-furnace systems without special considerations need to be in the 30,000-40,000-lb/hr range to be attractive at $3 per 1,000 lb. (Remember to verify steam values thoroughly.) Special situations can be economical at any time, particularly if improvements in the over-all steam balance can be combined with the recovery project. With

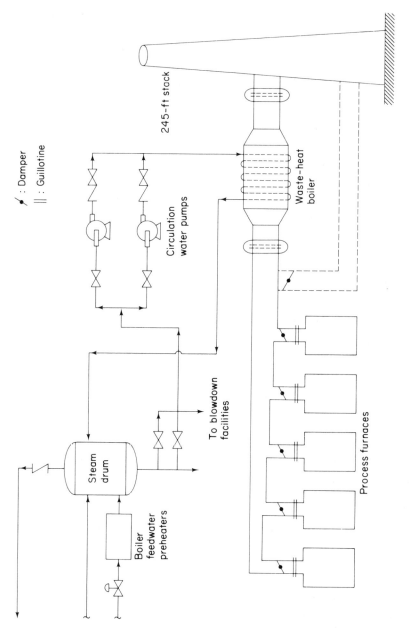

Figure 8-27. Heat recovery, case 1: natural-draft waste-heat boiler.

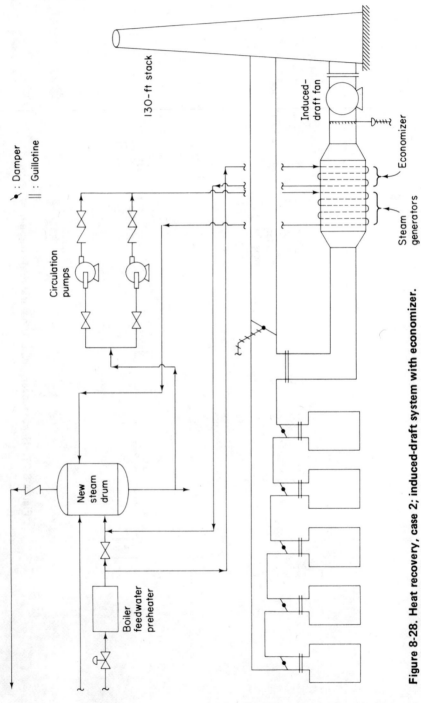

Figure 8-28. Heat recovery, case 2; induced-draft system with economizer.

Table 8-11. Multifurnace waste-heat-boiler study.

	Case 1	Case 2
Stack height (ft)	275	130
Duct velocity (ft/sec)	25	25
Draft	Natural	Induced
Fan horsepower	—	165
Boiler pressure (in. H_2O)	0.5	7.5
Steam capacity (lb/hr)	38,000	44,500
Economizer	No	Yes
Feedwater temperature (°F)	267	303
Stack temperature (°F)	450	350

Table 8-12. Waste-heat-boiler economics: Is it good?

	Steam Production (Thousands of Pounds per Hour)		
	< 10	30–40	> 40
Existing draft, convection-section additions	Probably	Yes	Yes
Retrofit of complete systems	No	Marginal	Yes
Special situations on flue gas	Possibly	Probably	Yes
Non-flue-gas process heat-recovery systems	Some	Yes	Yes

the breakdown given, each plant can develop its own guidelines based on its own steam values and capital-cost indices.

Waste-heat boilers on unfired process equipment are generally feasible at even smaller sizes than those for convection-section additions. Capital costs are lower because large flue-gas ducts and fans are not required. On the other hand, the range of steam pressures possible is often restricted to lower values by the available process temperatures. Again, the long-term value of the steam recovered must be verified carefully in developing economics for these projects.

Problem 8-5

1. Comment on the relative economics of air preheaters and waste-heat boiler systems. What factors detract from the advantages listed?

2. Consider the following list of tower condensers rejecting heat to cooling water:

	Duty (millions Btu/hr)	Process Temperatures (°F)	
		In	Out
E-12	15	350	200
E-201	23	375	350
E-5	17	300	290
E-7B	19	350	180

These units are located in a single process block. All towers are reboiled by a hot-oil belt. The hot oil is supplied by a large furnace with a stack temperature of 400°F, which uses ambient combustion air at 20 percent excess. Hot oil leaves the furnace at 550°F and returns at 350°F. What possibilities would you suggest for energy recovery in this system?

3. Your plant has a large calcining operation which fires natural gas at high excess air to remove the internal water from an inorganic crystalline product. A schematic of the operation is shown. You anticipate a natural-gas curtailment this winter of 30 percent, but you do not wish to cut production. How might this be achieved? What order of magnitude investment will be required?

Answers

1. Two factors are apparent from the examples presented on waste-heat boilers and air preheaters. Under normal circumstances more heat can be recovered and less capital is required for air preheat systems. In addition, credits are surer, since fuel is conserved directly. The value of the heat saved does not depend on the overall site steam balance.

Several factors detract from these advantages. Steam is more flexible and far more transportable than large combustion characteristics and the radiant-heat distribution of a furnace and can increase emissions of nitrogen oxides. The addition of hot-air ducts can sometimes cause access problems under a furnace and may in fact become uneconomical for furnaces with a large number of wall burners.

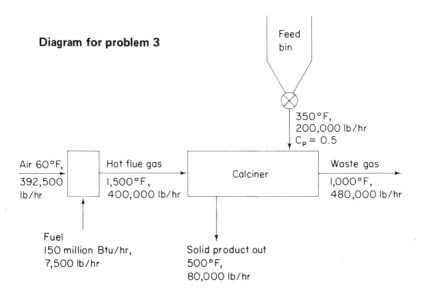

Diagram for problem 3

2. In this problem the furnace is relatively efficient (82 percent from the available-heat diagram in unit 1), and the condensers, taken individually, are only medium-sized and not high-temperature. Taken collectively, the total heat rejected is 74 million Btu/hr and that fired in the furnace is about 90 million Btu/hr. To accomplish this about 4,500 lb/hr of fuel is required, using about 90,000 lb/hr air. From 70°F, air preheat duty requirements are

	Air Temperature			
	100°F	200°F	250°F	300°F
Duty (million Btu/hr)	0.65	2.8	3.9	5.0

Thus, air preheat can provide only a small use for the condenser heat, since most of its duty is required at temperatures higher than those at which the heat is being rejected.

A waste-heat boiler is an economic possibility if the heat can be dealt with collectively. This idea leads directly to the concept of a hot belt to substitute for the cooling water, as shown in the diagram. Pressurized water would be the first choice because the temperature drop for an intermediate fluid is eliminated. Since about 50°F is required between source and steam temperature, the maximum steam pressure possible would be about 60 psig. The diagram gives an approximate schematic of the proposed system. Until the details of the specific temperature—heat transfer curves for the condensers are known, only order-of-magnitude balances are possible.

Diagram for solution 2

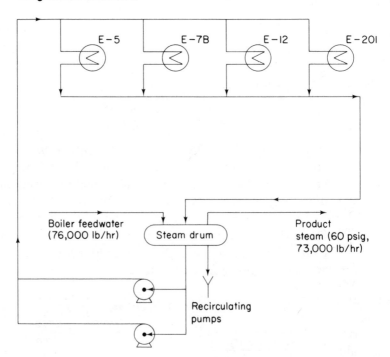

The capital cost of the system will involve additional heat exchange as well as hot-belt piping and a steam drum. Temperature differences in each exchanger will be reduced and more heat-exchanger shells will be required. The cooling-water exchangers may still be required to complete condensation. Since relatively low-pressure water is being circulated, the mechanical design of the exchangers may be adequate.

The economics of the system cannot be determined from the available data. They will depend strongly on the value for low-pressure steam in this plant and on the extent of new facilities required.

To enhance the value of heat recovered, if needed, a steam compressor could be considered.

3. The calciner problem presents a classic choice between air preheat and waste-heat boilers. We must keep in mind, however, that the ultimate goal is to reduce consumption of natural gas in the face of shortage. In the present case only about 50 million Btu/hr are being absorbed into the process.

Feed heating:	$200,000 \times 0.5 \times (500 - 250)°F$	= 25*
Water vaporization:	$20,000 \times 970$	= 19.4*
Water heating:	$20,000 \times 0.5 \times (1,000 - 500)$	= 5.0*
Total		49.4*

[*millions of Btu per hour]

Since 150 million Btu/hr are being fired, chances are good that fuel savings of 30 percent are possible.

The performance of a waste-heat boiler can be estimated from the gas-turbine data given in Table 8-10. Since most duties are at relatively high temperature, high steam pressures are of interest. About 52,000 lb/hr of 630 psig steam could be generated. This could be used to preheat the feed and perhaps do some of the water vaporization, but this amounts to replacing the calciner—a high-capital-cost route.

The steam could also be used to preheat the combustion air to about 450°F. This would utilize 400,000 x 0.24 x (450 − 60) = 37.4 million Btu/hr—not enough for the target saving, and about twice the surface area of a direct air-preheat system. If high pressure steam were being produced from natural-gas fuel elsewhere in the plant, this route might be attractive.

Direct flue-gas air preheat will cool the flue gas about one degree for each degree that it heats the fresh air. Preheating the air to 560°F will cool the flue gas to about 500°F, and no condensation should occur. The duty of the preheater would be

$$400,000 \text{ x } 0.24 \text{ } (560 - 60) = 48 \text{ million Btu/hr},$$

or about the desired saving.

Very roughly, the capital cost would be about $5 million. Simple fuel savings at $6 per million Btu would be about $2.4 million/yr for around-the-clock operation; therefore, the project should be economical. Of course, if the alternative is curtailed production and loss of sales, much greater incentives exist.

Beyond the air-preheater case one might well be able to improve this system even further. The exhaust flue gas from the air preheater could be used to generate steam or to preheat feed (in a direct-contact fluid bed, for example), and the product solids might also be put to use. Either holds the potential for another 20 million Btu/hr savings if a use can be found for the recovered heat.

GLOSSARY

availability
> The maximum amount of work that could be extracted from a system changing from an initial state to a state in mutual stable equilibrium with the environment.

combustion air
> The amount of air injected to a burner to support the combustion reaction. All this air must be heated to combustion temperature.

combustion-air preheat
> Use of energy derived from sources other than the flame to supply part of the heat necessary to heat the air to combustion temperature.

draft
> The lower pressure produced in a furnace stack by the temperature difference between the hot flue gas and the cooler ambient air outside the stack.

entropy
> A property of state. An increase in entropy represents a decrease in the quality, usefulness, or availability of the energy content of the system.

excess air
> The amount of air supplied to a combustion process over the stoichiometric value. This is normally expressed as a percentage excess, that is 10 percent excess air means 110 percent of the stoichiometric need.

flue gas
> The products of combustion of fuel going up the stack.

heat pump
> An engine that transfers energy from a low temperature to a high temperature by using work.

isentropic
> Occurring with no change in entropy.

latent heat

The energy change when a substance changes phase at constant temperature; for example, the energy given off when a pound of steam condenses at constant temperature.

stack temperature

The temperature of flue gas as it enters the stack.

stoichiometric amount

The amount of a reactant required to complete a given chemical reaction. The stoichiometric amount of combustion air would be the amount needed to convert all fuel to carbon dioxide and water.

throttle valve

A valve used to reduce the pressure of a fluid without energy recovery.

About the Author

William F. Kenney is Manager of Facilities Improvement in the Olefins Technology Division of the Exxon Chemical Company. Previously, he initiated and supervised the Energy Conservation Technology Section for all of Exxon Chemical. He also gained extensive process-design and project-management experience at Allied Chemical and at Brookhaven National Laboratory. He holds a BSChE from Yale (1955) and an MSChE from Purdue University (1956), and is a member of the American Institute of Chemical Engineers.

9

Heat Recovery
Case Studies and Examples

CASE STUDY 9-1
USE ENGINE EXHAUST HEAT
TO GENERATE STEAM

By replacing the conventional muffler on a 500 hp gas engine with a waste heat recovery muffler at a cost of $60,000, a process plant is able to realize an annual savings of about 55,600 gallons of fuel oil costing $69,500.

The engine, which is used to drive a compressor, discharges 4225 cfm of exhaust at 1100F and 6 inches of water back pressure. The plant requires 2000 lb/hr of 15 psig steam, which was supplied by a fired boiler using No. 2 fuel oil, and has available 220F water for boiler feedwater. The exhaust heat recovered by the waste heat recovery muffler can provide approximately 1300 lb/hr of the required steam allowing the output of steam from the fired boiler to be reduced resulting in the substantial fuel oil savings. The calculations are shown below.

Calculations

1. First, we estimate the heat recovery from the exhaust gas when cooled from 1100F to 350F, which is 100F above the temperature of 15 psig saturated steam.

We enter Figure 9-1 with the initial and final exhaust gas temperatures, 1100F and 350F respectively, and determine the approximate heat recovery to be 198 Btu/lb exhaust gas.

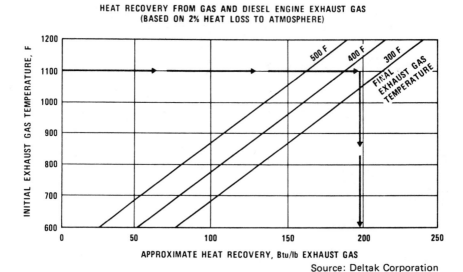

HEAT RECOVERY FROM GAS AND DIESEL ENGINE EXHAUST GAS
(BASED ON 2% HEAT LOSS TO ATMOSPHERE)

Source: Deltak Corporation

Figure 9-1. Heat Recovery from Gas and Diesel Engine Exhaust Gas
(Based on 2% Heat Loss to Atmosphere)

Knowing that the specific volume of the flue gas* at 1100F is 39.5 cu ft/lb, we convert the exhaust flow rate from 4225 cfm to lb/h as follows:

$$\text{Exhaust flow rate} = \frac{4225 \text{ cu ft/min} \times 60 \text{ min/h}}{39.5 \text{ cu ft/lb}}$$

$$= 6420 \text{ lb/h}$$

Then, rate of heat recovery = 6420 lb/h × 198 Btu/lb
= 1.27 MBtu/h

2. Next, we calculate the amount of steam generated from the exhaust heat recovered. From Figure 9-2 we determine that with

_____
*The specific volume of flue gas at a given temperature is very close to the specific volume of air at that temperature.

HEAT ABSORBED IN GENERATION OF SATURATED STEAM

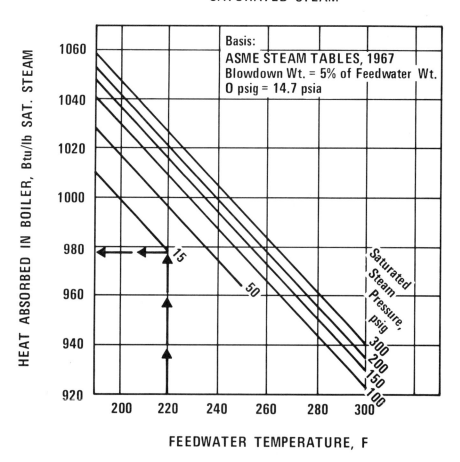

Figure 9-2. Heat Absorbed in Generation of Saturated Steam

220F feedwater the heat absorbed in generating 25 psig saturated steam is 977 Btu/lb steam.

$$\text{Steam generation} = \frac{1.27 \text{ MBtu/lb}}{977 \ \text{ Btu/lb}} = 1300 \text{ lb/h}$$

Thus, steam generation from the fired boiler is reduced from 2000 lb/h to 700 lb/h.

3. The resulting annual fuel savings with 80% boiler efficiency based on the lower heating value of the No. 2 fuel oil (18,300 Btu/lb oil) and assuming 4500 hours of operation per year are calculated below:

$$\text{Annual fuel savings} = \frac{1.27 \text{ MBtu/h} \times 4500 \text{ h/yr}}{0.8}$$

$$= 7140 \text{ MBtu/yr}$$

Knowing that the fuel oil density is about 7.02 lb/gal., we figure the annual fuel savings in gallons.

$$\text{Annual fuel savings} = \frac{7140 \text{ MBtu/yr}}{18,300 \text{ Btu/lb} \times 7.02 \text{ lb/gal.}}$$

$$= 55,600 \text{ gal./yr}$$

4. If the No. 2 fuel oil costs $1.25 $/gal.

$$\text{Annual savings} = 55,600 \text{ gal./yr} \times 1.25 \text{ $/gal.}$$

$$= \$69,500 \text{ per year}$$

The total cost of a waste heat recovery muffler and installation is about $60,000 for a 500-hp engine. In this case, the installation can be paid off in about one year.

CASE STUDY 9-2
RECOVER BOILER FLUE GAS HEAT
FOR SPACE HEATING AND
FEEDWATER PREHEATING

An opportunity for energy conservation was identified through the recovery of heat loss in the exhaust gases from a natural gas-fired boiler of a dairy. Additional natural gas was not available for a conventional expansion of the boiler room capacity to heat a warehouse expansion. Heat would be recovered by a finned coil installed in the boiler stack breeching to transfer heat from hot flue gases to circulating water. This hot water would be pumped to fan coils within the warehouse and to a boiler feedwater preheater. When more heat is available than needed, flue gases would be automatically diverted to the atmosphere.

1. Heat Required for Warehouse Addition

$$q_1 = UA dt$$

where q = heat flow, Btu/h
   U = heat transfer coefficient, Btu/h sq ft $°F$
   A = area, sq ft
   dt = difference in indoor and mean outdoor temperature
     = 50F for boiler
   $i = 1, 2, 3$

Heat loss through walls

Lower portion, brick and block, U = 0.33 Btu/h sq ft $°F$,
   8 ft high, 371 ft exposed perimeter

q_1 = 0.33 Btu/h sq ft $°F$ × (371 ft × 8 ft) × 50F
   = 49,000 Btu/h

Upper portion, U = 0.25 Btu/h sq ft $°F$, 12 ft high, 371 ft exposed perimeter

q_2 = 0.25 Btu/h sq ft $°F$ × (371 ft × 12 ft) × 50F
   = 55,700 Btu/h

Heat loss through roof

Flat, insulated 1½ in. fiberboard plus tar and gravel,
   U = 0.22 Btu/h sq ft $°F$, 135 ft × 101 ft

q_3 = 0.22 Btu/h sq ft $°F$ × (135 ft × 101 ft) × 50F
   = 150,000 Btu/h

Heat loss in ventilating air exhaust

Air changes required = 0.5 warehouse volume per hour

$$q = W_a C_p dt$$

where q = heat flow, Btu/h
   W_a = air flow, lb/h

C_p = specific heat of air = 0.24 Btu/lb °F
dt = difference in temperature of indoor and outdoor air
= 50F

$$q_4 = \frac{0.5 \times 135 \text{ ft} \times 101 \text{ ft} \times 20 \text{ ft}}{h}$$

$$\times \frac{0.075 \text{ lb}}{\text{cu ft}} \times 0.24 \frac{\text{Btu}}{\text{lb °F}} \times 50\text{F}$$

= 123,000 Btu/h

Total hourly heat for warehouse

$$= q_1 + q_2 + q_3 + q_4$$

= 49,000 Btu/h + 55,700 Btu/h + 150,000 Btu/h
+ 123,000 Btu/h

= 378,000 Btu/h

Total annual heat for warehouse

= 378,000 Btu/h × 24 h/day × 6,000 degree days/year
× 1/50F

= 1,090 MBtu/yr

2. Heat Required to Raise Feedwater Temperature 50F

Fuel gas flow rate = 50,000 kcf/yr
Fuel gas heating value = 1 MBtu/kcf
Boiler efficiency = 75%
Enthalpy of steam minus enthalpy of feedwater
= 1,185 Btu/lb
Steam flow rate

$$= \frac{50,000 \text{ kcf/yr} \times 1 \text{ MBtu/kcf}}{1,185 \text{ Btu/lb}} \times 0.75$$

= 31.6 Mlb/yr

Disregarding boiler blowdown, the feedwater flow rate is equal
to the steam flow rate. Using a specific heat for water of 1 Btu/lb °F,

Feedwater preheat required = 31.6 Mlb/yr $\times$ 1 Btu/lb °F
$\times$ 50F = 1,580 MBtu/yr

3. Total Heat Requirement = 1,090 MBtu/yr + 1,580 MBtu/yr
= 2,670 MBtu/yr

4. Heat available from flue gas

Twenty-five percent of the heat of combustion of the fuel gas is lost in the flue gas. Considering that it is possible to recover 75% of the heat loss in the stack,

Heat available from flue gas
= 50,000 kcf/yr $\times$ 1 MBtu/kcf $\times$ 0.75 $\times$ 0.25
= 9,380 MBtu per year

Heat available for future use
= 9,380 MBtu/yr $-$ 2,670 MBtu/yr
= 6,710 MBtu per year

However, any future heat requirement would have to be a warm weather application since all recoverable heat will be used by this system in very cold weather.

5. Annual Savings

The savings realized by heating the warehouse with the heat recovery system would be equivalent to the cost of the alternative of electric heating.

Cost of electric heating of the warehouse addition:

Additional energy rate   = $.08/kwh
Additional demand charge = $10/kw month

Additional electric energy
 cost $= \dfrac{1,090 \text{ MBtu/yr}}{3,412 \text{ Btu/kwh}} \times \$.08/\text{kwh}$

$= \$25,556$

Monthly demand charge $= \dfrac{378,000 \text{ Btu/h}}{3,412 \text{ Btu/kwh}} \times 10 \text{ \$/kw month}$

$= \$1,107.85 \text{ per month}$

Using this charge for Dec., Jan., Feb.:

$$\$1,107.85 \times 3 = \$3,323.55$$

Using one-half this charge for Oct., Nov., Mar.:

$$\frac{1}{2} (\$1,107.85) \times 3 = \$1,661.62$$

Annual demand charge   = $4,985/yr

Total electric heat cost   = $30,541/yr

Using the heat recovery system for feedwater heating saves natural gas costing $4/mcf.

Annual cost of gas for feedwater heating (at 75% efficiency)

$$= 1,580 \text{ MBtu/yr} \times 4 \text{ \$/MBtu} \times 1/0.75$$
$$= \$8,426.66$$

Total savings for space heating and feedwater preheating

$$= \$3,480/yr + \$8,426.66/yr = \$5,360 \text{ per year}$$

6. Capital Costs and Payback Period

The following costs are based on estimates from local contractors and will vary depending on equipment details and local prices.

Cost of heat recovery system for
warehouse and feedwater heating   = $43,350

Cost of alternate electric heating system
for warehouse   = $20,750

Incremental first cost for heat recovery
system   = $64,100

$$\text{Payback period} = \frac{\$64,100}{\$30,541/yr} = 2.09 \text{ years}$$

CASE STUDY 9-3
RETURN STEAM CONDENSATE
TO BOILER PLANT

In a plant where the value of steam is $6/MBtu, saturated steam was delivered to one building at a pressure of 200 psig, at an average rate of 27,000 lb/h, and for an average of 8,000 h/yr. The steam was reduced through control valves and condensed in heating coils at an average pressure of 25 psig. The condensate was returned to the boiler plant and used as feedwater. The amount and value of the heat recovered is calculated below.

From Figure 9-3 below, it is determined that 17 percent of the heat remains in the 25 psig condensate from the 200 psig saturated steam.

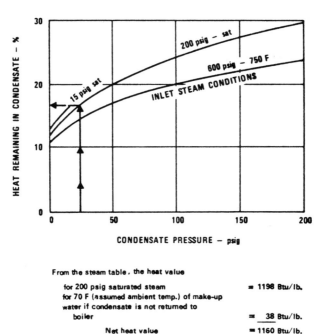

From the steam table, the heat value

for 200 psig saturated steam	= 1198 Btu/lb.
for 70 F (assumed ambient temp.) of make-up water if condensate is not returned to boiler	= 38 Btu/lb.
Net heat value	= 1160 Btu/lb.

Figure 9-3. Heat in Steam Condensate
(calculated from steam tables)

The heat recovered in condensate

$$= 17\%/100\% \times 1{,}160 \text{ Btu/lb} \times 27{,}000 \text{ lb/h} \times 8{,}000 \text{ h/yr}$$
$$= 42{,}600 \text{ MBtu per year}$$

The value of the heat recovered

$$= 42{,}600 \text{ MBtu/yr} \times \$6/\text{MBtu} = \$255{,}600$$

It should be noted that these values represent heat saving potential since no heat loss has been considered for returning the condensate to the boiler plant. The heat loss is dependent on factors such as the length of return lines and the amount of insulation.

In addition to saving heat, the return of condensate to the boiler plant will:

1. Save treated makeup boiler feed water.

2. Save energy and chemicals used in the water treating operation.

3. Reduce water pollution.

4. Reduce (but not eliminate) the cost of losses due to steam trap leakage.

CASE STUDY 9-4
WASTE HEAT RECOVERY IN
AN ASPHALT ROOFING PLANT

Background

A plant which produced asphalt for impregnated felt roofing, was required, for pollution control purposes, to dispose properly of fumes from the asphalt-flowing still and the felt saturator. Two methods were considered. One was the use of a scrubber; the other was the use of a fume incinerator to eliminate the odor. Since incineration is a high temperature oxidation process and the heating value of the fumes was low and variable, some fuel was needed to maintain the required temperature in the incinerator.

The plant also used process steam, supplied by four boilers which were about 20 to 40 years old. The plant manager realized that the energy in the fumes, together with that in the natural gas used for the incinerator, could be recovered by the addition of a heat recovery boiler at the end of the incinerator, thus reducing the amount of fuel used for process steam generation.

Specifications for Design of the Waste Heat Recovery Device

One waste heat boiler was to be designed to take heat from the flue gas from an incinerator burning fumes. The unit was to be a skid-mounted, shop-assembled, watertube boiler to continuously generate 15,000 lb/h of steam at 150 lb/in.^{2} g pressure.

The following was the process information for the boiler:

(1) Waste Gases
<table>
<tr><td>Volume</td><td>16,000 scfm</td></tr>
<tr><td>Inlet temperature</td><td>1400F</td></tr>
<tr><td>Outlet temperature</td><td>600F approx.</td></tr>
<tr><td>Pressure drop</td><td>6 in. of water max.</td></tr>
<tr><td>Fouling factor</td><td>0.001 h · ft^2 · F/Btu</td></tr>
</table>

(2) Boiler Feedwater and Steam
<table>
<tr><td>Feedwater pressure</td><td>175 lb/in.^{2} g</td></tr>
<tr><td>Feedwater temperature</td><td>200F</td></tr>
<tr><td>Steam quantity (saturated)</td><td>15,000 lb/h</td></tr>
<tr><td>Steam side fouling factor</td><td>0.001 h · ft^2 · F/Btu</td></tr>
</table>

The unit was to be strictly designed and stamped to section I of the ASME Boiler and Pressure Vessel Code. The boiler was to be of watertube, natural circulation design. Extended surfaces, such as finned tubes, were acceptable. Carbon steel tubes (SA-178A) and shell plates (SA-515-70) were to be used. The following accessories were to be included:

(a) Standard boiler trim, such as safety valves, water column with gage glasses and try cocks, steam pressure gage, feed and check valves, continuous blowdown valves and intermittent blow-off valves.

(b) A single-element feedwater regulator with by-pass valve.

(c) A steam stop-check valve.

The unit should have horizontal gas inlet and vertical gas outlet.

The manufacturer was to provide, with his proposal, information on the heat surface and overall heat transfer coefficient, the calculated gas pressure drop, the gas outlet temperature, and the steam purity in ppm of solid carryover in the steam.

Description of Waste Heat Boiler Design

One water tube waste heat boiler was designed for this application. The unit generated 15,000 lb/h of saturated steam at 150 lb/in.^{2}g. The thermal and mechanical data are shown in Table 9-1. The general outline of the unit is shown on Figure 9-4.

Table 9-1. Performance and Data

PURCHASER:		DATE:		DES.:
FOR:		TYPE:		PROJ.:
PLANT SITE:		DRAWING NO.:		DRAWN:
PURCHASERS REFERENCE:		NUMBER		EST.:

UNIT					
	Gas flow	lb/h	73,300		
	Design pressure/temperature	in/F	20		
	Operating pressure at inlet	in			
	Operating pressure at outlet	in			
	Pressure drop	in	3.5		
	Temperature at inlet	F	1,400		
	Temperature at outlet	F	628		
	Temperature drop	F	772		
	Specific heat (ave.)	Btu/lb/F	.28		
	Heat given up	MBtu/h	15.85		
	Efficiency	%	98		
	Fouling factor	hr, sq ft, F/Btu	.001		
	Auxiliary fuel (lhv =) lb/h				
	Fluid flow	lb/h	15,000		
	Design pressure/temperature	psig/F	200		
	Operating pressure at inlet	psig	150		
	Operating pressure at outlet	psig	150		
	Pressure drop	psi	—		
	Temperature at outlet	F	366		
	Temperature at inlet	F	200		
	Temperature rise	F	166		
	Heat absorbed/lb of steam	Btu/lb	1,036		
	Heat absorbed	MBtu/h	15.52		
	LMTD	F	562		
	Overall heat tr. coeff.	Btu/hr, sq ft, F	9.63		
	Fouling factor	hr, sq ft, F/Btu	.001		
	Steam quality	% or PPM	3		
	Blow down water	%	5		
	Heating surface	sq ft	2,970		
	Tube diameter	in	2		
	min. thickness	in	.105		
	length				
	material		SA-178A		
	No. of rows transverse to flow		28, 10F		
	Transverse spacing (ST)	in	3 3/4		
	No. of rows longitudinal to flow		5 (2 half rows bare)		
	Longitudinal spacing (SL)	in	4		
	Fin no/in		6		
	height	in	5/8		
	thickness	in	.05		
	material		CS		
	Drum (header) diameter	in	42.24		
	length	in			
	material	in	SA-515-70		
	Inlet nozzle diameter	in	14		
	Outlet nozzle diameter	in	4		

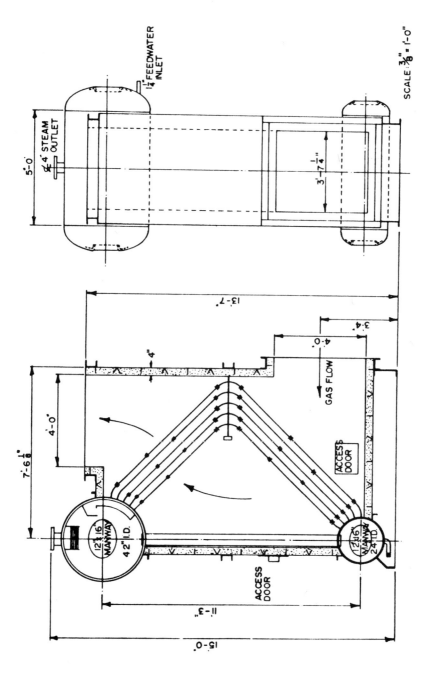

Figure 9-4. Waste Heat Boiler

The heating surface of the boiler consisted of bare tubes and finned tubes. The first two rows of the first pass were bare tubes and were radiant shield tubes for the boiler, since the incinerator was to be hooked up to the boiler and the tubes were exposed to the flames. The remaining tubes were finned.

The unit was completely shop-assembled. It was designed, built and stamped according to section I of the ASME Boiler and Pressure Vessel Code. The gas flow arrangement had a horizontal inlet and vertical outlet. The short stack was mounted on the top of the boiler.

Cost Evaluation of the Pollution Control and Waste Heat Recovery System

One way of analyzing the cost of this investment is as follows:

Fume collecting hood, ductwork, and forced draft fan	$ 22,500
Incinerator	62,500
Waste heat boiler	80,000
Stacks	5,000
Piping for steam line	5,000
Field installation	75,000
Total Cost	$250,000

In this case, the total benefit from the waste heat boiler was the provision of 15,000 lb/h of 150 lb/in.^{2}g steam which formerly had to be purchased at a cost of $6.00/1000 lb. This is equivalent to $270,000/yr. The additional cost of installing the waste heat boiler was approximately $80,000 which gives a payback period of approximately 4 months.

Review of the Application

The incinerator was installed primarily because of pollution control regulations. However, process steam was needed in the plant and it was advantageous to add the waste heat recovery boiler to the incinerator. Although the waste heat boiler had about the same efficiency as the fired boiler, less fuel was used to generate the same amount of steam because of the energy recovered from the fumes.

The waste heat boiler was designed to satisfy the steam requirement in this specification, not for the maximum heat recovery. Therefore, there was additional sensible heat remaining in the stack flue gas which could have been recovered by an economizer on the top of the boiler if more steam had been needed.

CASE STUDY 9-5
HEAT RECOVERY BY GAS TO GAS
REGENERATION IN A
HYDROCARBON FUME INCINERATOR

Background

A potential air pollution problem was introduced when another production can line was to be added at a large can company's plant. Hydrocarbons were emitted from solvents utilized in the bake process following lithographic printing of coated steel sheets for caps and closures. At times the solvent loadings encountered in the effluent were quite high; at other times they were considerably lower. Therefore, an outside contractor was retained to engineer, fabricate, deliver and install an incineration system which could include heat recovery for economy while being sensitive to the variable nature of the effluent.

Description of Previously
Existing Equipment

The new can line was the third one installed at the plant site. The two previously existing lines had fume incineration; but those incinerators did not include provision for any type of heat recovery from the products of incineration. The process can be schematically represented by going directly from the forced draft fan of Figure 9-5 to the combustor without preheat, followed by direct discharge of combustion exhaust as clean, uncooled gas to the stack.

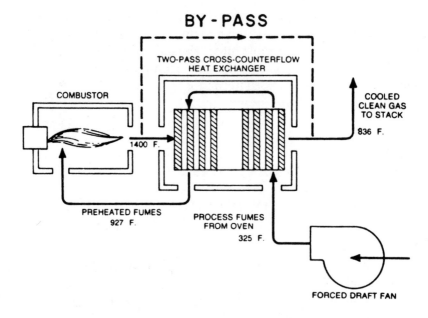

Figure 9-5. Schematic Diagram of Fume Incineration System Used in
Lithographic Printing of Coated Steel Sheets
Heat Exchanger is Used to Recover Heat
and Reduce Fuel Cost in Incineration.

Specifications for Design of the
Waste Heat Recovery System

The primary specifications for the waste heat and incineration system included:

1. complete destruction of hydrocarbons in the gases from the bake ovens;

2. direct-flame incineration;

3. operational economy.

Specifically, the first of these objectives was recognized to be the most restrictive requirement, because the high solvent loadings encountered at times in the effluent had a correspondingly high heat of combustion. At times negligible heat release results from the actual incineration of the fumes, but at other times more heat may be released than can be handled without special provisions in the system.

Description of the Fume Incineration System

A schematic of the fume incineration system is shown as Figure 9-5. Oven exhaust fumes were directed from the process gas forced draft fan through the inside of the tubes of a recuperative type heat exchanger where they were preheated before entering the combustor. This heat exchanger was a two-pass cross-counterflow system with gases from the bake ovens entering at up to 325F and leaving at up to 927F to enter the combustor.

In the combustion chamber, the fumes were exposed under proper conditions to a direct flame which oxidized the contaminants to carbon dioxide and water vapor.

The incineration chamber was designed so that the incinerated gases remain in the chamber for a minimum residence time of 0.3 seconds to insure the complete destruction of the hydrocarbons in the gases.

The now-incinerated hot gases (about 1400F) left the direct-flame combustion chamber and passed over the outside of the heat exchanger tubes, going to the stack as cleaned, cooled gases at about 836F. Of course, this cooling was accomplished by heat energy transfer from the clean gases to the contaminated incoming oven exhaust fumes.

The system was made up of modular components for a wide range of performance capability and arrangement flexibility. The heat exchanger had a hot gas bypass control (discussed more later), an incineration chamber complete with burner, safeguard controls and instrumentation, a forced-draft process gas fan, platform and support steel for rooftop mounting, ductwork, dampers, expansion joints and a stainless steel stack with rain cap protection.

Cost Study of the
Recovery System

The total cost of the incineration-heat recovery system, including installation and research work, was $300,000. [The fuel cost per year for a process flow rate of 8500 scfm with heat recovery was estimated at $170,000 against $370,000 without recovery (54% savings), when operating at an incineration temperature of 1400F.] This was based on 6,000 h/yr of operation and a fuel cost of $5/ MBtu. The calorific value of the effluent was not considered because of the widely varying solvent loadings.

A simple payback period calculation can now be made. Based on the savings in fuel cost per year of $200,000 and total first cost of $300,000, one readily finds the payback period to be 1.5 years.

Difficulties Encountered in Design
and Special Features in the System

To accommodate the very high solvent loadings encountered in the effluent, a hot gas side heat exchanger bypass was provided (Figure 9-5). This bypass automatically opens when solvent heat release by combustion is unusually high, thereby lowering the temperature of the "preheated" fumes entering the combustor. Thus, this damper functioned to maintain a minimum burner fuel requirement, restrict the maximum operating temperature, and achieve good incineration efficiency.

A raw gas burner was provided because of its operational economy. This type of burner requires no primary combustion air, but utilizes available oxygen from the air in the effluent stream to support combustion.

Comparison of Normal Operation Results
Before and After the Recuperation
Incineration Installation

A major benefit from the installation of the recuperation incineration system as compared to operation of the bake ovens without either incineration or recuperation was clean exhaust gases.

While incineration without recuperation (as for the two previously existing can lines) also would have led to reduction of a potential odor and pollution problem, it would have cost more in fuel consumption.

One significant later addition to the system was the installation of baffle plates in a dead corner of the burner section to better direct the air flow. Under consideration is the addition of a catalytic cone-type converter between the burner and heat exchanger. It is anticipated that this would cut the entrance temperature to the heat exchanger from the present 1400F to about 700F, which could result in 40–45 percent additional reduction in fuel consumption.

CASE STUDY 9-6
HEAT RECLAMATION FOR
HVAC SYSTEM[1]

Background

This case study describes an energy engineering project in which a business engaged and worked with a consultant on an energy conservation equipment retrofit project. Described are the detailed steps involved in successfully completing such a project from inception through startup, including the time required and the responsibilities of both the business and the consultant.

In this project the business was Western Products in Milwaukee, Wisconsin and the consultant was Affiliated Engineers, Inc. of Madison, Wisconsin. The project was an HVAC expansion featuring heat reclamation.

Problem Definition

Facility Description. Western Products' facility is a 120,000 sq ft, two-floor building in Milwaukee. The building, originally 34,000 sq ft in 1965, was expanded in 1968 and in 1974. Remodeling was done in 1976.

The facility houses offices, a computer room, warehouse space and production facilities, including welding, painting and general machining. The product manufactured is a steel vehicle-mounted snowplow.

Problem Identification. By 1977, the HVAC system in the facility had become inadequate. Ventilation in offices, welding, and machine areas was insufficient. The building had a significant negative pressure, due to the high exhaust rates necessary for welding areas. The HVAC system was undersized, leading to noticeable loss of comfort during cold winter periods.

Normally, such difficulties would be handled by expansion of the existing HVAC system to provide more heated ventilation make-up air. However, the winter heating costs were already up to $5,000 per month in the winter of 1976–77. More importantly, the local natural gas supplier, facing shortages, refused to provide expanded service to industrial customers, including Western Products. These concerns, together with large heat losses resulting from high exhaust rates, suggested a potential heat recovery system application.

Identifying and Evaluating Options

Selection of a Consultant. Western Products had limited experience in HVAC modifications or in energy conservation and reclamation equipment. They therefore decided to obtain professional help from an outside consultant. In selecting the consultant, Western Products based their evaluations primarily on the professional qualifications of the firms under consideration. Once the firm best suited to execute the project was identified, negotiations were conducted to determine compensation.

Relevant qualifications considered included: (1) the firm's size, relative to the size of the proposed project, (2) the firm's experience, both overall and in similar projects, (3) the qualifications and professional registration of the firm's staff, and (4) references from the firm's current and/or former clients.

The firm's proximity to the client was not overemphasized. An experienced firm at a remote location from the client is usually cost competitive with a less experienced local firm.

As a result of their consultant search and evaluation, Western Products contracted with Affiliated Engineers in January of 1977.

Site Survey. As a first step in identifying and evaluating options, the consultant carried out a site survey. The purpose of the survey was twofold; it allowed members of the consultant's staff to become familiar with the facility, and it provided an opportunity for the consultant to begin to identify potential solutions to the problem. The survey was conducted by a team of engineers who had experience with similar applications.

Physical data, such as dimensions, temperatures, and air flows were obtained. The condition of the existing equipment was assessed. Discussions were held with operating personnel to identify deficiencies and desired additional features. Potential options were identified, and other data necessary for their evaluation was obtained.

Evaluation of Options. After the site survey and discussions with the owner, the next step was to evaluate the possible options in depth. The four options identified for evaluation were:

1. Add new ductwork and additional cooling for offices and computer room.

2. Add a new makeup air unit, with heat reclamation in the machine shop area.

3. Improve the exhaust of smoke in the welding shop by increasing air flow. This option also includes heat reclamation.

4. Add a machine shop ventilating unit and rebalance the supply air to correct building pressurization.

The results of the evaluations were presented in a report, in which the estimated construction cost of each alternative was presented within ±20%. A generic description of each option was also presented. The report then summarized the results and the consultant's recommendations. The report was dated April 1977.

Recommendations. After examining the benefits and costs of the various options, the owner then determined the most appropriate course to follow. These actions were as follows:

1. Add ductwork and fans to improve air distribution in machine shop, welding shop, and office space.

2. Add heat reclamation on the increased air flow supply in the welding shop.

3. Balance the air flow in the facility to set the building at a 1% positive air supply.

These solutions represent a combination of elements of each of the consultant's recommendations. They were arrived at by the owner after consideration of both desired system improvements and cost limitations.

Design of Problem Solution

Once the solution of the problem was determined qualitatively, the client then instructed the consultant to prepare construction documents. These documents were to be used to obtain competitive bids for construction of the solution.

Design Criteria. The first step in this process was establishing design criteria. Design criteria form the basis upon which the design is executed, and include both subjective and objective elements, such as relative quality of construction, design life, type and supplier of equipment to be used, and future space functional changes. Input from the owner was critical in establishing these criteria, and included specific requirements of the system and its components, as well as any future changes in facility layout.

In establishing the design criteria for the type of heat reclaim equipment to be used, the consultant evaluated three alternative types of equipment. They were: (1) heat recovery wheels, (2) heat pipe heat exchangers, and (3) air-to-air heat exchangers.

The owner indicated that the exhaust air in the welding shop was extremely dirty. This precluded the use of a heat recovery wheel, because the dirty exhaust air stream would plug the media in the wheel, which would result in low effectiveness and high maintenance costs. An economic comparison of the air-to-air unit with the heat pipe unit showed that the heat pipe unit was more cost effective because of its higher effectiveness and smaller space requirement. As a result, the heat pipe unit was selected as the basis for the design.

The owner also indicated that the welding shop would not be cooled in the summer, and that fresh air ventilation was highly desirable. Therefore in the summer a necessary design criterion was that a bypass around the heat pipe unit be provided to prevent overheating of the space. The heat pipe unit is unused during the summer. Therefore, variable tilt devices, which allow the heat pipes to reverse heat flow direction and precool outside air for air conditioning, were necessary.

Construction Documents. Based on the design criteria, the consultant then developed the construction documents for the project. These included six drawings and a 30-page specification, which described the required equipment and installation in detail.

The specification was divided into various areas of concern for each piece or type of equipment, or for certain procedures. The equipment specification sections were further divided in three subsections: General, Products, and Execution.

The General section covers testing, shop drawing submittals and details, and related comments. The Products section covers the specifics of the manufacture and operation of the equipment. The Execution section details the equipment installation.

The wording of the specification was as precise and as explicit as possible. To serve as an example, the following paragraphs are excerpted from the Products subsection of the heat pipe unit specification.

FABRICATION

Casings on coil shall be fabricated from No. 16 gauge galvanized steel, of a design proven to mount and support such coils in ductwork or in a piece of apparatus. All casings shall be furnished with a minimum of 1½" wide flange on each of the four sides, both front and back for mounting. Bolt slots for mounting shall be punched on 3" centers throughout. Intermediate core supports shall be furnished to prevent sagging or bowing when mounted in the recommended manner. Coil to have aluminum non-corrugated plate fin with 1" tube.

Coil shall be complete with drip pan.

Working fluid shall be commercially pure group 1 refrigerants as classified in the American National Standard Safety Code for Mechanical Refrigeration. Prior to charging at the factory, each tube shall be cleaned, dried and evacuated.

Protective end covers shall be provided to protect the tube ends which extend through the casing. They are to be constructed of No. 20 gauge galvanized steel and will be factory sealed.

Total coil size shall be 144 inches long by 108 inches high. The coil bank shall be comprised of two 144 inch by 54 inch coils. Fresh air side shall be 6 rows, 14 fins per inch. The exhaust air side shall be 6 rows, 11 fins per inch. The coil bank shall have a total free area of 108 sq ft. The coils shall be capable of handling 26000 cfm of exhaust air at 60°F and 50% RH, and 24000 cfm of fresh air at —15°F up to a maximum of 140°F.

Entire unit must be capable of withstanding entrained moisture from outside air. If required, unit must withstand washdown of cleaning without harm to operation.

Bids. Once the construction documents were completed, the consultant assisted the owner in identifying qualified and experienced contractors. A complete set of construction documents was then sent to each of the four approved contractors, as well as an invitation for bids on July 22, 1977. Bids were received and tabulated by August 4, 1977. Bids were for the total construction of the project and differed by approximately 25% from high to low. In this case, after evaluation by the owner and the consultant, the low bid was accepted.

Construction. Once the contractor was selected, a construction schedule was established by the consultant, owner, and contractor. Construction was carried out without interruption of the facility's operation and was completed with a final air balancing on February 27, 1978, slightly later than originally anticipated.

Results

Economics of Heat Reclamation. In this section, the owner's additional investment in energy conservation equipment included only the heat pipe unit, and some additional ductwork (Figure 9-6). The remainder of the system components were necessary to provide improved ventilation. Therefore, the economic practicality of the investment in energy savings equipment depended only on the cost of installing the heat pipe unit.

The design conditions specified for the heat pipe unit are —15F outside ambient and 60F indoor temperature. Under these extreme conditions, the 14,000 cfm of makeup air would be preheated from —15F to 31F, representing a heat reclamation rate of

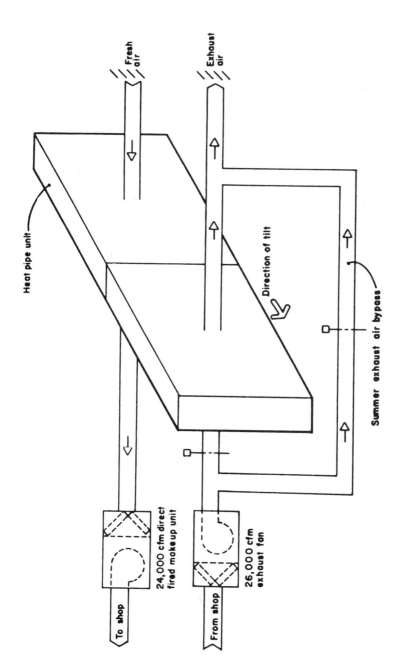

Figure 9-6. Heat Reclaim System Schematic

1.192 MMBtus/hr. However, under more typical outdoor conditions, heat reclamation, as well as overall building heating requirements, will be lower.

The average ambient temperature during the 30 week heating season in Milwaukee is 32.9F. Assuming the ambient to be a constant of 32.9F with 60F exhaust air, the heat pipe unit will save 2,124 MMBtus per heating season.

A simplified economic analysis of the investment is as follows:

Installed Costs (1978)	
Heat Pipe Unit	$18,130
Extra Ductwork	469
Cost for Larger Fan	754
Total First Cost (Installed)	$19,353

These values include labor, supervision, contractor's overhead and profit, and taxes.

The yearly savings are calculated as:

Energy Savings	2,124 MMBtu/yr
Equivalent Dollar Savings	
(Natural Gas @ $0.22/therm)	$4,673
Maintenance of Reclaim Coil	
(12 hr/yr @ $14.28/hr)	($171/yr)
Net Savings	$4,502/yr

The simple payback period is therefore calculated as:

$$\frac{\$19,353}{\$\ 4,502} = 4.3 \text{ yrs}$$

Operational Experience. The system has been in operation for 3 years and modifications have been made. In the original design, warm and dirty exhaust air passed through 20% filters before entering the heat pipe unit. As clogging of the coil was occurring, the general contractor changed to an 80% filter. A filter of such density is generally only recommended for hospitals. In this industrial situation, these filters clogged within days. Also, this large increase in filter density so restricted exhaust air flow that frosting occurred on the exhaust coil.

At about the same time as these filter changes were occurring, the plant engineer who supervised the installation of the heat pipe unit for the owner left his job. His replacement had no previous experience with heat reclamation systems.

Subsequently, because the maintenance staff was not familiar with the function or design of the unit, they neglected to correct the filter problem and avoided servicing it. The unit was not cleaned for an extended period of time. The maintenance staff was bypassing exhaust air around the unit throughout the year, rendering it ineffective.

These problems were solved by a brief on-site meeting between the owner and the consultant. The proper filter type was identified. The maintenance staff, and management, were instructed by the consultant on proper maintenance procedures, and proper operation of the unit was restored.

Summary

The key steps in the project described in this paper are typical of many energy retrofit projects. The project began with the identification of the problem by the owner in 1976. The owner realized that he did not have the capability to solve his problem in-house, and obtained assistance from an outside consultant. The consultant then performed a site survey, evaluated the options and made recommendations on the most cost effective option, by April 1977. Based on the owner's directions, construction drawings and specifications were developed and issued by the consultant. A contractor was selected, by competitive bids, in August of 1977. Construction was completed in February of 1978. Follow-up work on the project has consisted of answering questions on proper system operation.

Several of the lessons learned in this project apply to similar projects. For example, in projects where construction is to be carried out without interruption of normal space functions, construction schedules should contain some allowances for unexpected delays.

Also, the operational difficulties evident in this project point up the need to both communicate maintenance tasks to the owner's maintenance staff and to document maintenance needs in writing. The owner should maintain up-to-date maintenance manuals for reference and for new employees, and should conduct a preventative maintenance program.

The significance of this project is that it demonstrates the ability of a heat reclamation unit to save money, and more importantly, to solve a utility source problem. Without heat reclamation, the owner could not have improved ventilation in his facility. The local natural gas supplier was not allowing any expansion of service, and other sources were far less economical. Heat reclamation served as a new energy source to heat larger quantities of makeup air. In addition, the heat pipe unit recovered its own installation cost in 4.3 years of energy cost savings.

The following table summarizes the major steps in this project, the elapsed time required for each, and the areas of responsibility for both the consultant and the owner.

Activity	Time Required to Complete	Owner's Activities	Consultant's Activities
(1) Definition of problem		X	
(2) Site survey	1 month	X	X
(3) Evaluation of options and recommendations	2 months		X
(4) Decision by owner on course of action and establishment of design criteria	2 weeks	X	X
(5) Design of solution	3 months		X
(6) Bids	1 week	X	X
(7) Construction	7 months	X	X

CASE STUDY 9-7
RECOVERY OF WASTE HEAT
FROM A PROPELLANT
FORCED AIR DRY HOUSE

Background

This case study[2] is a description of the installation and evaluation of a waste heat recovery system designed to recover heat from a dry house used for drying multi-base propellants. Recovery is accomplished by transferring heat from contaminated hot air being exhausted from a dry house to preheat fresh ambient air being supplied to the house. Descriptions of the special operating and safety features of the waste heat recovery unit are also presented. Evaluation of the system showed that up to 62% of the energy used to dry multi-base propellants can be saved by the technique employed. Energy/cost saving projections are shown which indicate that approximately $1,500,000 per year can be saved at a major propellant manufacturing facility if this waste heat recovery technique were fully implemented.

Introduction

The Army has a network of government-owned, contractor-operated munitions plants scattered all over the country. These Army Ammunition Plants are responsible for the manufacture of a wide variety of propellants, explosives, and metal parts, and the loading of these components into a wide variety of ammunition items. Even though these plants are not operating at maximum capacity they consume large amounts of energy—a total of close to 20 trillion Btus annually. Because of the growing concern in the Army community that energy in appropriate quantities might not be available to meet future mobilization requirements at these plants, a comprehensive energy conservation technology program was established in 1976 to accomplish the dual goal of: (1) reducing energy consumption, and thereby reducing the cost of munitions manufacturing and loading, and (2) conserving our dwindling fossil fuel resources. The program that was created is a multi-task effort which is organized to: (1) survey the energy consumption on a unit operation/process basis, (2) introduce new energy conservation technology that will result in reduced energy consumption per unit of product manufactured, and (3) adapt alternate sources of energy to plant operations in order to reduce dependence on fossil fuels.

One of the tasks under this program is "Energy Recovery from Waste Heat." This section discusses one of the conservation projects recently completed under this task; namely, the recovery of heat from the hot air exhausted to the atmosphere during the drying of multi-base propellant (propellant containing nitrocellulose plus an energetic liquid plasticizer, usually nitroglycerin, plus one or more additional energetic solid ingredients). This application represents a typical example of a process operation at an Army plant where large amounts of energy can be saved through effective heat transfer technology. It also illustrates one of the special problems that is frequently encountered in applying energy conservation technology to munitions operations; the involvement with explosives and/or explosive contaminated environments.

Present Method of Operation

The present method for drying multi-base propellant is to force heated air through a dry house containing solvent-wet propellant stacked in drying trays. Typical solvents that may be present in the propellant during the drying stage are ethyl alcohol, acetone, and ethyl ether. Figure 9-7 is a schematic drawing showing the general layout of two typical forced air dry houses (FADs) in combination with one fan house which supplies air to both dry houses. Outside air is heated by steam coils to 140F and then forced through the dry house once. As the warm air circulates through the stacked trays of solvent-wet propellant it absorbs the solvents along with trace quantities of nitroglycerin before being exhausted to the atmosphere through a vent close to the bottom of one wall of the dry house (Figure 9-8). Immediately outside the exhaust vent is a canopy which serves to condense the nitroglycerin which drops into a catch-pan directly below the vent. The catch-pan contains an inerting solution which desensitizes the nitroglycerin. This is a single-pass process with the exhaust air temperature nearly equal to the warm dry house inlet air temperature. Until now, no effort has been made to reclaim this wasted heat, which can amount to 9 or 10 million pounds of steam per year for each dry house in operation at a propellant manufacturing facility such as Radford Army Ammunition Plant.

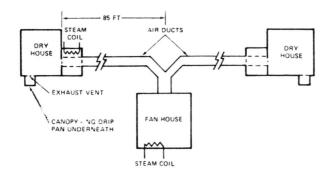

Figure 9-7. Normal Dry House Operation at ARRADCOM

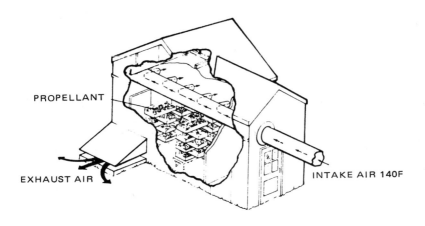

Figure 9-8. Multi-Base Propellant Dry House

Selection of a Heat Exchanger

By using an air-to-air heat exchanger between the exhaust and intake streams, large amounts of the waste heat exhausting from the dry house can be used to preheat the cold intake air. This results in a significant reduction in the energy demands for the propellant drying process and important cost and energy savings to the Army.

Although there are three other competitive air-to-air energy recovery devices commercially available, a heat pipe heat exchanger (HPHX) was selected as the most responsive to the performance requirements of the forced-air dry house application.

Rotary regenerative units (i.e., thermal wheels) were eliminated because cross-stream contamination is permitted by their design (rotation of heat transfer surfaces between adjacent air streams). At best, this leakage can amount to 10% of the air flow. In addition, relatively high maintenance is associated with these units' seals, drive, and motor.

Plate-fin heat exchangers are generally of cross-flow design which limits their theoretical effectiveness to 75% as opposed to 100% for counterflow units.

Counterflow designs have more complicated ducting arrangements since both intake and exhaust flows are turned 180°. Since sealing between air streams is required over the entire area of thin sheet-metal heat transfer surfaces, long-term corrosion can also be a problem.

Finally, the run-around-loop, which uses an intermediate heat transfer fluid, does not lend itself to this application, being better suited to widely separated air streams where side-by-side ducting would be impractical. It also requires electric power and added maintenance to accommodate the pumped fluid loop.

A schematic drawing of a typical heat pipe is shown in Figure 9-9. These devices essentially consist of a sealed, partially evacuated metal tube with a special wick material along its inner wall and a small quantity of working fluid. When heat is applied to one end of the heat pipe, called the evaporator, the fluid is vaporized and travels up the center of the tube to the cooler end where it condenses, loses its latent heat, and returns along the wick by capillary action, aided by gravity, to the heated end to complete the cycle.

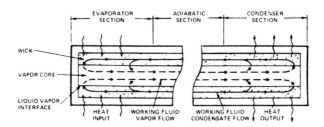

Figure 9-9. Internal Configuration of a Heat Pipe

Performance Advantages of Heat Pipe Heat Exchanger

The performance advantages of a HPHX are summarized below:

High Effectiveness. These units have an effectiveness of up to 80% with simple fin-type configurations.

Simple Control Over Effectiveness. Wide variations in effectiveness can be achieved by either mechanically tilting the unit slightly or by diverting a portion of the air flow around the exchanger.

Separation of Inlet and Exhaust Air Streams. A sealed center partition prevents cross-stream infiltration and contamination.

Easy Maintenance and Cleaning of Heat Transfer Surfaces. Spray nozzles and access ports are readily accommodated for cleaning and maintenance of the unit.

Main Features of Waste Heat Recovery System

Figure 9-10 shows a schematic drawing of the layout of the waste heat recovery system installed on a propellant dry house at the US Army Armament Research and Development Command (ARRADCOM) at Picatinny Arsenal in Dover, New Jersey. An isometric drawing of the system is shown in Figure 9-11.

Heat Pipe Heat Exchanger. Details of the HPHX are presented in Table 9-2. The all-aluminum unit contains 180 individually sealed heat pipe tubes that are mechanically press fitted to 12-mil-thick plate fins using conventional heat exchanger design practice. The heat pipe working fluid is Freon-12, chosen because it has a higher

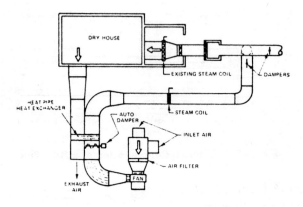

Figure 9-10. Schematic Diagram of Waste Heat Recovery System

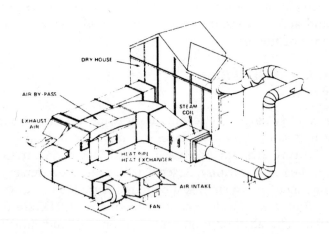

Figure 9-11. Isometric Drawing of Waste Heat Recovery System

Table 9-2. Heat Exchanger Features

Aluminum Heat Pipe and Fins	
Exchanger Dimensions:	Length—69 inches
	Height—31 inches
	Width—12 inches
	Area per Side—6.25 sq ft
Heat Pipe Dimensions:	Length—64 inches
	O.D.—0.625 inch
	Wall Thickness—0.035 inch
180 Heat Pipes:	9 Rows
	20 Heat Pipes per Row
	12 Fins per inch
Freon-12 Working Fluid	

than atmospheric vapor pressure at all anticipated service temperatures (0–200F) and is nonreactive with the exhaust stream contaminants in the event of a tube leak.

For this application, a fin spacing of 12 fins per inch was considered the closest spacing consistent with reasonable maintenance practices in a clean air stream environment (nonparticulate). A standard tube pattern consisting of 5/8-inch OD tubes arranged in staggered rows and placed on 1.5-inch centers was also judged acceptable. The applicable performance variation at a 3000 cfm flow rate as a function of face velocity (V_F) and number of tube rows is summarized in Figure 9-12.

The final heat exchanger size (i.e., frontal area, number of rows) was determined by selecting the configuration that best combines a large net energy savings and a high rate of return on investment. For instance, using the highest possible effectiveness unit increases exhaust stream energy recovery but requires a larger, more expensive unit with more heat pipes and/or closer finning, which results in larger pressure drops and greatly increased fan power requirements. Increased fan power decreases the net energy savings of the system, and combined with the increased heat exchanger costs, also decreases the rate of return. Based on the results of the return on investment

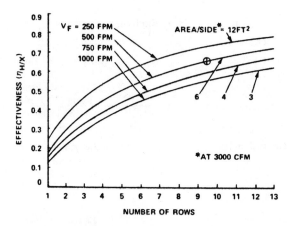

Figure 9-12. Heat Pipe Heat Exchanger Performance

(ROI) analysis (Figure 9-13) a final heat exchanger design having a nominal frontal area of 6 sq ft per side and 9 rows of heat pipes (20 heat pipes per row) was selected. The nominal heat exchanger effectiveness is 65% for balanced intake and exhaust design flow rates of 3000 cfm.

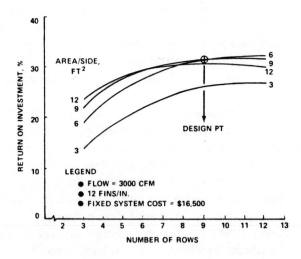

Figure 9-13. Return on Investment vs. HPHX Size

Automatic Inlet Air Bypass Control. The system is designed to automatically prevent the nitroglycerin in the exhaust stream from condensing until after it leaves the heat exchanger. This requires the temperature of the exhaust stream to be maintained above the nitroglycerin dew point of 80F through the exchanger. This is accomplished by a temperature control system which consists of: (1) a bypass duct over the top of the heat exchanger on the air inlet side, (2) a set of dampers in the bypass duct and the through duct to the exchanger on the air inlet side, (3) pneumatic damper actuators, and (4) a pneumatic temperature controller. A schematic drawing showing the duct bypass configuration is presented in Figure 9-14.

As the exhaust air temperature approaches the dew point of nitroglycerin, the bypass dampers progressively open while the in-line dampers partially close. This reduces the amount of heat removed from the exhaust air stream and maintains its temperature above the nitroglycerin dew point. If a system malfunction permits the exhaust air temperature to drop below the nitroglycerin dew point, an alarm sounds and the air inlet fan automatically shuts down.

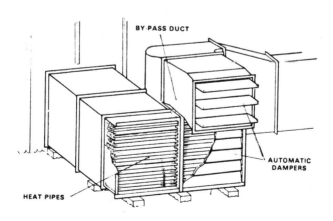

Figure 9-14. **Heat Pipe Heat Exchanger**

Push Through Fan. The centrifugal fan circulates 3000 cfm with a static pressure rise of 3 inches water. The fan is made from non-sparking materials and is driven through an adjustable V-belt drive by a 3-hp, 460-volt 3-phase 60-Hz electric motor. The 18.25-inch diameter impeller wheel has a nominal speed of 1500 rpm. One of the critical requirements of the system design is to prevent infiltration of the clean inlet air from the nitroglycerin and solvent contaminated exhaust stream. This is accomplished by creating a sealed partition between opposing air ducts complimented by a push through fan location. Locating the fan upstream of the heat exchanger on the air inlet side permits the air pressure of the inlet air stream through the exchanger to exceed that of the exhaust air stream. Thus, in the remote event that leakage were to occur across the center partition of the heat exchanger, it would flow from the clean air stream to the contaminated exhaust air stream and not vice-versa.

Temperature Controlled Steam Coil. The waste heat recovery system also includes the installation of a 300-pound-per-hour steam coil in the inlet air duct between the heat exchanger and the dry house. A pneumatic signal from a temperature sensor in the inlet duct between the exchanger and the steam coil modulates a pressure reducing valve which regulates steam flow to the coil and raises the inlet air temperature to the dry house level of 140F. A secondary steam coil in the inlet air duct just before the dry house tempers the air to 140F when needed. The system also includes a high temperature safety feature which automatically sounds an alarm and shuts down the fan if the air temperature in the inlet duct exceeds 180F.

Evaluation of Waste Heat Recovery System

Test Description. The main test objective was to obtain a quantitative comparison of the total energy consumption for a typical forced-air dry house operating with and without the HPHX waste heat recovery system. Additionally, data were obtained on the operation of the automatic intake bypass control system so that its performance benefits could be weighed against the increased system complexity and cost.

The dry cycle sequence was representative of that used for M30 propellant at Radford Army Ammunition Plant. It consisted of an 11-hour warmup from a standby condition at an average temperature rise of 5F per hour, followed by a 40-hour drying cycle at 140F during which the heat exchanger hot side exhaust temperature was limited to 80F. The M30 propellant load for each dry cycle was 2500 pounds.

Test Set-up/Instrumentation. For each test sequence, the energy required to operate the dry house was determined by monitoring the total air stream energy input. This was obtained by measuring the differential temperatures and air flow associated with each heat transfer component (heat exchanger, steam coils).

The volumetric flow rate of air entering the dry house was obtained by measuring the air velocity at the centerline of the inlet duct, about 20 feet downstream of the nearest bend. A careful probe of the duct cross-section determined that the measured centerline velocity was very close to the average duct velocity, giving a difference of less than ½%. The product of the centerline duct velocity and duct cross-sectional area (2.88 sq ft) gives the volumetric flow rate (cfm).

A kilowatt-hour meter, part of the fan electrical installation, was used to measure HPHX fan power consumption. The total reading was manually recorded prior to and immediately after each data-taking period. Except for the fan power consumption, all data were automatically recorded at regular time intervals (typically 20 minutes) by a multi-channel data logger which produced a permanent paper tape record of all transducer measurements.

Test Results. Results from the evaluation of the waste heat recovery unit are summarized in Table 9-3. Two separate trials were conducted. The tests were performed in winter conditions and showed that the inlet air was raised an average of about 46F through the heat exchanger. Based upon these data, it was projected that the yearly recovery effectiveness would approximate 57%. If the design were altered to accommodate condensation of the nitroglycerin in the heat exchanger and subsequent removal in a manner similar to current practice, the recovery effectiveness would be increased to 62%.

Table 9-3. Performance of Waste Heat Recovery Unit

$T_R\downarrow$	$\uparrow T_4$
EXHAUST	INTAKE
$T_n\downarrow$	$\uparrow T_1$

DRY CYCLE	1	2
AIR FLOW (CFM)	3013	2914
DRY HOUSE TEMP ($^\circ$F)	140.7	139.7
INTAKE TEMP ($^\circ$F)		
T_1	23.4	25.8
T_4	69.1	72.4
EXHAUST TEMP ($^\circ$F)		
T_R	138.6	138.7
T_n	85.9	83.9
BTUs SAVED CYCLE	$64 \cdot 10^6$	$62 \cdot 10^6$
EFFICIENCY (WINTER EVALUATION) (WITHOUT NG CONDENSATION)	42	44
PROJECTED YEARLY EFFICIENCY		
WITHOUT NG CONDENSATION	57	
WITH NG CONDENSATION	62	

Economic Analysis. Economic analyses were also performed on the system based upon potential implementation at Radford Army Ammunition Plant, one of the Army's major propellant manufacturing facilities. The analyses were conducted at two levels; implementation to meet the current production rate and implementation to meet maximum production capacity, during mobilization conditions. Results from these analyses are shown in Table 9-4. They indicate that 46.5 billion Btu per year can be saved from implementation of this waste heat recovery technique to meet current production and about 220 billion Btu per year under mobilization conditions. This is equivalent to an average annual savings over a 10-year life cycle of $306,000 at current levels and $1.47 million at mobilization.

In each instance payback is projected to be reached in 1.7 years.

Table 9-4. **Economic Analysis of Waste Heat Recovery System at Radford Army Ammunition Plant**

● APPLICATION: PREHEATING OF INLET AIR
● PROPOSED METHOD: USING HEAT FROM EXHAUST AIR
● AIR FLOW: 5500 CFM (PER BAY)
● AIR TEMPERATURE: 140° F

	CURRENT PRODUCTION	MOBILIZATION
POTENTIAL ENERGY SAVINGS—BTU/YEAR	46.5×10^9	220×10^9
SYSTEM COST — $1000	518	2,486
YEARLY SAVINGS — $1000	306	1,471
PAYBACK PERIOD — YEARS	1.7	1.7
YEARLY ENERGY EQUIVALENT — SCF GAS	$46.5\overline{M}$	$220\overline{M}$
GAL OIL	332,000	$1.6\overline{M}$

Summary

A demonstration waste heat recovery unit using a heat pipe heat exchanger as the heat transfer vehicle has been successfully installed and operated on a multi-base propellant forced-air dry house at the US Army Armament Research and Development Command in Dover, NJ. Evaluation of the system showed that up to 62% of the waste heat discharged from the dry house can be recovered to preheat clean incoming air without compromising safety. Economic projections show that implementation of the recovery technique at Radford Army Ammunition Plant will yield savings as high as $1.47 million per year.

Acknowledgment

Grumman Aerospace Corporation of Bethpage, NY was responsible for the design, fabrication, installation and evaluation of the heat pipe heat exchanger under contract to ARRADCOM.

REFERENCES

1.   J. C. Nelson and W. A. Ryan, "The Practical Application of Heat Recovery Concepts," *Advances in Energy Utilization Technology,* Proceedings of the 4th World Energy Engineering Congress. The Fairmont Press, Inc., Atlanta, GA 30324.

2.   J. M. Swotinsky, E. F. Bozza, and P. Mullaney, "Recovery of Waste Heat from a Propellant Forced Air Dry House," *Energy Management,* Proceedings of the 3rd World Energy Engineering Congress. The Fairmont Press, Inc., Atlanta, GA 30324.

Index